Great Science Adventures

The World of Plants

by
Dinah Zike
and
Susan Simpson

Common Sense Press

See where learning takes you.

www.greatscienceadventures.com

Great Science Adventures is a comprehensive series which is projected to include the titles below. Please check our website, www.greatscienceadventures.com, for updates and product availability.

Great Life Science Studies:
    The World of Plants
    The World of Insects and Arachnids
    The World of the Human Body
    The World of Vertebrates
    The World of Biomes
    The World of Health

Great Physical Science Studies:
    The World of Tools and Technology
    The World of Matter and Energy
    The World of Light and Sound
    The World of Electricity and Magnets

Great Earth Science Studies:
    The World of Space
    The World of Atmosphere and Weather
    The World of Lithosphere / Earth
    The World of Hydrosphere / Fresh Water
    The World of Hydrosphere / Oceans
    The World of Rocks and Minerals

Great Science Adventures

Table of Contents

Great Science Adventures

Introduction

Great Science Adventures is a unique, highly effective program that is easy to use for teachers and students. This book contains 24 lessons. Concepts being taught are clearly listed at the top of each lesson. Activities questions, clear directions, and pictures are included to make the work easy for you and your students. Each lesson will take one to three days to complete.

This program utilizes highly effective methods of learning. Students not only gain knowledge of basic science concepts, but also learn how to apply them.

Specially designed *3D Graphic Organizers* are implemented in the lessons. These organizers review the science concepts while adding to your students' understanding and retention of the subject matter.

This *Great Science Adventures* book is divided into four parts.

Following this *Introduction* you will find the *How to Use this Program* section. It contains all the information you need to make the program successful. The *How to Use* section also contains instructions for Dinah Zike's *3D Graphic Organizers* used in the program. Please take the time to learn the terms and instructions for these learning manipulatives.

In the *Teacher's Section,* a number is used to designate each lesson. The lessons include a list of the science concepts to be taught, simple to complex vocabulary words, and activities that reinforce the science concepts. Each activity includes a list of materials needed, directions, pictures, questions, written assignments, and other helpful information for the teacher.

The *Teacher's Section* includes enrichment activities entitled *Experiences, Investigations, and Research.* Alternative assessment suggestions are found at the end of the *Teacher's Section.*

The *Lots of Science Library Books* are next. These books are numbered to correlate with the lessons. Once completed, the *Lots of Science Library Books* contain all the content needed for each lesson. You may read the *LSLB* books to your students, ask them to read the books on their own, or make the books available as research materials. Covers for the books are found at the beginning of the *LSLB* section. (Common Sense Press grants permission for you to photocopy the *Lots of Science Library Books* pages and covers for all of your students.)

Graphic Pages, also listed by lesson numbers, provide pictures and graphics that can be used in the activities. They can be duplicated and used on student-made manipulatives or students may draw their own illustrations. At the front of this section are the Lab Log, *Investigative Loop*, and a world map that may be photocopied as well. (Common Sense Press grants permission for you to photocopy the *Graphics Pages* for all of your students.)

A Special Note about Plants

Since the 4th century, life forms have been divided into two kingdoms: plants and animals. As techniques for examining the cells of living things have improved, it is clear that the major divisions of the living world should extend beyond plants and animals. Today, divisions of living things are determined by the cellular structure of the organism.

This science program is based on the six-kingdom classification system of the living world. In this system, the plant kingdom is made up of multicellular organisms that usually have walled cells and contain chloroplasts where they produce food by photosynthesis. Algae, bacterium, and fungi are not included in the plant kingdom in this system of classification.

Below is a simple explanation of the six-kingdom classification system. Please refer to current research materials if you desire more information.

The six-kingdom classification includes:
1.   Kingdom Archaebacteria
2.   Kingdom Eubacteria
3.   Kingdom Protista
4.   Kingdom Fungi
5.   Kingdom Plantae
6.   Kingdom Animalia

How to Use This Program

This program can be used in a single level classroom, multilevel classroom, homeschool, co-op group, or science club. Everything you need for a complete plant study is included in this book. Intermediate students will need access to basic reference materials.

Take the time to read the entire *How to Use this Program* section and become familiar with the sections of this book described in the *Introduction*.

Begin each lesson by reading the *Teacher Pages* for that lesson. Choose the vocabulary words for each student and the activities to complete. Collect the materials you need for these activities.

Introduce the lesson with the *Lots of Science Library Book* by reading it aloud or asking a student to read it. (The *Lots of Science Library Books* are located after the *Teacher's Section* in this book.)

Discuss the concepts presented in the *Lots of Science Library Book,* focusing on the ones listed in your *Teacher Section.*

Follow the directions for the activities you have chosen.

How to Use the Multilevel Approach

The lessons in this book include basic content appropriate for grades K-8 at different mastery levels. For example, when learning photosynthesis, a first grader may master the concept that plants make their own food by using water, sunshine, and air. This student is exposed to more information but not expected to retain it. In the same lesson, a sixth grade student will learn all the steps of photosynthesis, be able to communicate the process in writing, and apply that information to different situations with plants.

In the *Lots of Science Library Books,* the words written in the larger type are for all students. The words in smaller type are for upper level students and include more scientific terminology, details about the basic content, and interesting facts for older learners.

In the activity sections, icons are used to designate the levels in specific writing assignments.

This icon ✎ indicates the Beginning level, which is the non-reading or early reading student. This level mainly applies to kindergarten and first grade students.

This icon ✎✎ is used for the Primary level. This includes the reading student who is still working to be a fluent reader. This level is primarily designed for second and third graders.

This icon ✎✎✎ denotes the Intermediate level, or fluent reader. This level of activities will usually apply to fourth through eighth grade students.

If you are working with a student in seventh or eighth grade, we recommend using the assignments for the Intermediate level, plus at least one *Experiences, Investigations, and Research* activity per lesson.

No matter what grade level your students are working on, use a level of written work that is appropriate for their reading and writing abilities. It is good for students to review data they already know, learn new data and concepts, and be exposed to advanced information and processes.

Vocabulary Words

Each lesson contains vocabulary words, used in the content of the lesson. Some of these words will be "too easy" for your students, some will be "too hard," and others will be "just right." The "too easy" words will be used automatically during independent writing assignments. Words that are "too hard" can be used during discussion times. Words that are "just right" can be studied by definition, usage, and spelling. Encourage your students to make these words their own in their writing and speaking.

Beginning students will use their vocabulary words to reinforce reading instruction, to enhance discussions about the topic, and to be copied in cooperative writing, or teacher guided writing.

Primary and Intermediate students can make a Vocabulary Book for new words. Instructions for making a Vocabulary Book are found on page 3. The Vocabulary Book will contain the word definitions and sentences composed by the student for each word. Students should also be expected to use their vocabulary words in discussions and independent writing assignments. A vocabulary word with an asterisk (*) next to it is designated for Intermediate students.

Using 3D Graphic Organizers

The *3D Graphic Organizers* provide a format for students of all levels to conceptualize, analyze, review, and apply the concepts of the lesson. The *3D Graphic Organizers* take complicated information and break it down into visual parts so students can better understand the concepts. Most *3D Graphic Organizers* involve writing about the subject matter. Although the content for the levels will generally be the same, assignments and expectations for the levels will vary.

Beginning students may dictate or copy one or two "clue" words about the topic. These students will use the written clues to verbally communicate the science concept. The teacher should provide various ways for the students to verbally restate the idea. This will reinforce the science concept and also encourage the students in their reading and higher order thinking skills.

Primary students may write or copy one or two "clue" words and a sentence about the topic. The teacher should encourage students to use vocabulary words when writing these sentences. As students read their sentences and discuss the topic, they will reinforce the science concept, increasing the students' fluency in reading, and higher order thinking skills.

Intermediate students may write several sentences or a paragraph about the topic. These students are also encouraged to use reference materials to expand their knowledge of the subject. As these tasks are fulfilled, students enhance their abilities to locate information, read for content, compose sentences and paragraphs, and increase vocabulary. Encourage these students to use the vocabulary words in a context that indicates understanding of the words' meanings.

Illustrations for the *3D Graphic Organizers* are found on the *Graphics Pages* and are labeled by the lesson number and a letter, such as 5-A. Your students may either use these graphics to draw their own pictures, or cut them out and glue them directly on their work.

Several of the *3D Graphic Organizers* expand over a series of lessons. For this reason, you will need a storage system for each students' *3D Graphic Organizers*. A pocket folder or a reclosable plastic bag works well. See page 1 for more information on storing materials.

Investigative Loop™

The *Investigative Loop* is used throughout *Great Science Adventures* to ensure that your labs are effective and practical. Labs provide context for the content of the science lessons. The lessons seem more real to the students and students, take ownership of the concepts, increasing understanding as well as retention.

The *Investigative Loop* can be used in any lab. The steps are easy to follow, user friendly, and flexible.

Each *Investigative Loop* begins with a **Question or Concept**. Use a question in this phase if the lab is designed to answer it. For example, the question could be: How does a plant respond to sunlight? The activity for this lab will show how a plant responds to sunlight, so a question is the best way to begin this *Investigative Loop*.

If the lab is designed to demonstrate a concept, use a concept statement in this phase, such as: Leaves release oxygen. The lab will demonstrate that fact to the students.

After the Question or Concept is formulated, the next phase of the *Investigative Loop* is **Research and/or Predictions.** Research gives students a foundation for the lab. By researching the question or concept, students enter the lab with a basis for understanding what they observe. Predictions are best used when the first phase is a question. Predictions can be in the form of a statement, a diagram, or a sequence of events. One example might be, "I think the roots will grow toward the water."

 The **Procedure** of the lab follows. This is a clear explanation of how to set up the lab and any ongoing tasks involved in it. Materials for the lab can be included in this section or may precede the entire *Investigative Loop.*

Whether the lab is designed to answer a question or demonstrate a concept, the students' **Observations** are of prime importance. Examining the lab in detail is usually the purpose of the lab. Instruct the students what they are to focus upon in their observations. The Observation phase will continue until the lab ends.

 Once observations are made, students must **Record the Data.** The data may be the changes observed in a lab component where diagrams or illustrations would be appropriate. Recording quantitative or qualitative observations of the lab is another important activity in this phase. Records may be kept daily for an extended lab or at the beginning and end for a short lab.

Conclusions and/or Applications are completed when the lab ends. Usually the data records will be reviewed before a conclusion can be drawn about the lab. Encourage the students to defend their conclusions with the data records. Applications are made by using the conclusions to generalize to other situations or by stating how to use the information in daily life.

 Next we must **Communicate the Conclusions,** an important part of any lab. This phase is an opportunity for students to be creative. Conclusions can be communicated through a graph, story, report, video, mock radio show, etc. Students may also participate in a group presentation.

In any lab, questions are asked as the activity proceeds. In the *Investigative Loop* these are called **Spark Questions.** Questions that the lab sparks in the minds of the students are important to follow and discuss when the lab ends. The lab itself will answer many of these questions, while others may lead to a new *Investigative Loop.* Assign someone to keep a list of all Spark Questions.

 One lab very naturally leads to another. This begins the Investigative Loop again. The phase called **New Loop** is a time to narrow down the lab to a new question or concept. It is brainstorming time for those involved in the new loop. When the lab has been decided upon, the *Investigative Loop* begins again with a new Question or Concept.

Take the time to teach your students to make qualitative and quantitative observations. Qualitative observations involve recording the color, texture, shape, smell, size (as small, medium, large), or any words that describe the qualities of the plant. Quantitative observations involve using a standard unit of measurement to determine the length, width, weight, mass or volume of the plant or plant parts.

Each student will make a Lab Book to store information about the Investigative Loops. The Lab Book is a Pocket Book. Instructions for making a Pocket Book are found on page 2. Your students will make new Lab Books as needed to glue side-by-side to the previous one. Complete instructions are in the *Teacher's Section.*

Predictions, data, and conclusions about the Investigative Loops are usually written on Lab Record Cards. These can be 3 x 5-inch index cards or paper cut to size.

When you begin an *Investigative Loop,* ask your students to glue or draw the graphic of the experiment on the pocket of the Lab Book. Each *Investigative Loop* is labeled with the lesson number and another number. These numbers are also found on the corresponding graphics. The completed Lab Record Cards will be labeled by Lab Number and placed in the appropriate pocket.

In the plant study, several Investigative Loops take days to complete. For this reason we have included a Lab Log to help you and your students keep track of which experiments to check each day. Simply make a note on the Lab Log for the number of days needed to complete the experiment using the Lab name or number. The Lab Log is found in the *Graphics Pages* and may be photocopied.

Each day, check the Lab Log to see which experiments should be worked on that day. You may wish to assign a student as Lab Log Secretary or give all students their own Lab Logs.

During an *Investigative Loop,* beginning students should be encouraged to discuss their answers to all experiment questions. By discussing the topic, the students will not only learn the science concepts and procedures, but will learn to organize their thinking in a manner which will assist them in later years of writing. This discussion time is very important for beginning students and should not be rushed.

After the discussion, work with the students to construct a sentence about the topic. Let them copy the sentence. These students can also write "clue" words to help them remember key points about the experiment and discuss it at a later time.

Primary students should be encouraged to verbalize their answers. By discussing the topic, the students will learn the science concepts, procedures, and learn to organize their thinking, thus increasing their ability to use higher level thinking skills. After the discussion, students can complete the assignment using simple phrases or sentences. Encourage students to share the information they have learned with others, such as parents or friends, thereby reinforcing all the content and skills covered in the lesson.

Even though Intermediate students can write the answers to the Lab Record Card assignments, the discussion process is very important and should not be skipped. By discussing the process of the experiments, students review the science concepts and procedures as well as organize their thinking for the writing assignments. This allows them to think at higher levels and thereby function at higher levels in their writing. These students should be encouraged to use their vocabulary words in their Lab Record Card writing assignments.

Ongoing Projects: Plant ID Book

In addition to the Lab Book discussed previously, your students will make a Plant Identification Book using Large Question and Answer Books. Instructions for making a Large Question and Answer Book are found on page 2. Large Question and Answer Books will be made as needed and glued side-by-side. Complete instructions are in the *Teacher's Section.*

Your students will collect or observe plants throughout the program and record them in the Plant Identification Book. A drawing of the plant will go on the tab, with information about the plant written under the tab.

This process will be greatly enhanced if students have access to a Nature Guidebook for your area. Determine with your students what information you consider important, research it in your Guidebook and write about it for each plant entry. Suggestions are found in the *Teacher's Section,* Lesson 1.

Experiences, Investigations, and Research

At the end of each lesson in the *Teacher's Section* is a category of activities entitled *Experiences, Investigations, and Research.* These activities extend concepts taught in the lesson, provide a foundation for further study of the content, or integrate the study with other disciplines. The following icons are used to notify you of the type of activity.

Life Science

Hands On

Geography

History

Literature

Math

Research

Cumulative Project

At the end of the program we recommend that students compile a Cumulative Project using the activities they have completed during their course of study. It may include the Investigative Loops, Lab Record Cards, and the *3D Graphic Organizers* on display.

Please do not overlook the Cumulative Project, as it provides immeasurable benefits for your students. Students will review all the content as they create the project. Each student will organize the material in his or her own unique way, thus providing an opportunity for authentic assessment and reinforcing the context in which it was learned. This project creates a format where students can make sense of the whole study in a way that cannot be accomplished otherwise.

Fast Food and Fast Folds

"If making the manipulatives takes up too much of your instructional time, they are not worth doing. They have to be made quickly, and they can be, if the students know exactly what is expected of them. Hamburgers, Hot Dogs, Tacos, and Shutter-folds can be produced by students, who in turn use these folds to make organizers and manipulatives." Dinah Zike

Every fold has two parts. The outside edge formed by a fold is called the **"mountain."** The inside of this edge is the **"valley."**

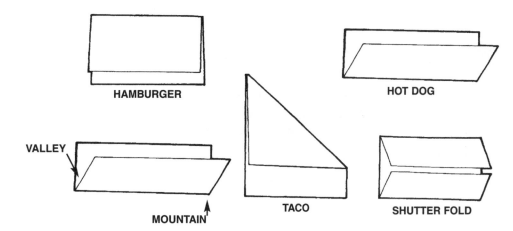

HAMBURGER

HOT DOG

VALLEY

MOUNTAIN

TACO

SHUTTER FOLD

Storage - Book Bags

One gallon reclosable plastic bags are ideal for storing ongoing projects and books students are writing and researching.

Use a strip of clear, 2" tape to secure 1" x 1" pieces of index card to the top corner of a bag under the closure, front and back. Punch a hole through the index cards and the bag. Use a giant notebook ring to keep several of the "Book Bags" together.

Label the bags by writing on them with a permanent marker.

The bags can be kept in a notebook by putting the 2" clear tape along the side of the storage bag, and punching 3 holes in the tape.

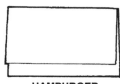

Half Book

1. Fold a sheet of paper in half like a Hamburger.

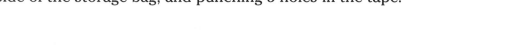

HAMBURGER

Large Question and Answer Book

1. Fold a sheet of paper in half like a Hamburger. Fold it in half again like a Hamburger. Cut up the valley of the inside fold, forming two tabs.

2. A larger book can be made by gluing Large Question and Answer Books "side by side."

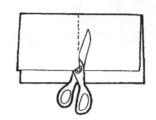

Small Question and Answer Book

1. Fold a sheet of paper in half like a Hot Dog.

2. Fold this long rectangle in half like a Hamburger.

3. Fold both ends back to touch the Mountain Top.

4. On the side forming two Valleys and one Mountain Top, make vertical cuts through one thickness of paper, forming tabs for questions and answers. These four tabs can also be cut in half making eight tabs.

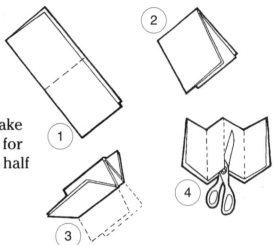

3 Tab Book

1. Fold a sheet of paper in half like a Hamburger or Hot Dog. Fold it into thirds. Cut up the inside folds to form three tabs.

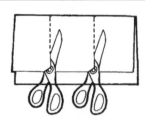

Pocket Book

1. Fold a sheet of paper in half like a Hamburger.

2. Open the folded paper and fold one of the long sides up three inches to form a pocket. Refold along the Hamburger fold so that the newly formed pockets are on the inside.

3. Glue the outer edges of the three inch pocket with a small amount of glue.

4. Make a multi-paged booklet by gluing several Pocket Books "side-by-side."

5. Glue a construction paper cover around the multi-page pocket booklet.

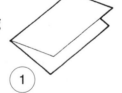
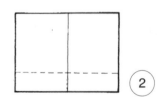
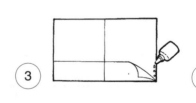
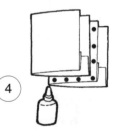

Vocabulary Book

1. Take two sheets of paper and fold each sheet like a Hot Dog.

2. Fold each Hot Dog in half like a Hamburger. Fold the Hamburger in half two more times and crease well. Open up the fold, and the sheet of paper will be divided into 1/16's.

3. On one side only, cut up the folds to the Mountain Top, forming eight tabs. Repeat this process on the second sheet of paper.

4. Take a sheet of construction paper and fold like a Hot Dog. Glue the solid back side of one vocabulary sheet to one of the inside sections of the construction paper. Glue the second vocabulary sheet to the other side of the construction paper fold. (This step can be eliminated to form a one sided vocabulary book.) Make sure the center folds of the vocabulary books meet at the center fold of the construction paper.

5. Vocabulary Books can be made larger by gluing them "side-by-side."

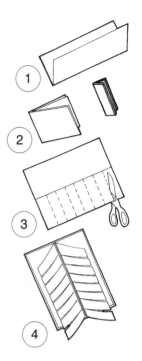

Pyramid Project

1. Fold a sheet of paper into a Taco.
   Cut off the excess tab formed by the fold.

2. Open the folded Taco and refold it the opposite way forming another Taco and an X fold pattern.

3. Cut up one of the folds to the center of the X and stop. This forms two triangular shaped flaps.

4. Glue one of the flaps under the other flap, forming a pyramid.

5. Place Pyramid on one end to make a diorama.

Layered Look Book

1. Stack two sheets of paper and place the back sheet one inch higher than the front sheet.

2. Bring the bottom of both sheets upward and align the edges so that all of the layers or tabs are the same distance apart.

3. When all tabs are an equal distance apart, fold the papers and crease well.

4. Open the papers and glue them together along the Valley/center fold.

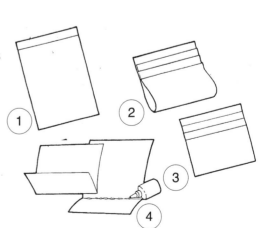

Trifold Book

1. Fold a sheet of paper into thirds.

2. Use this book as is, or cut into shapes. If the trifold is cut, leave plenty of fold on both sides of the designed shape, so the book will open and close in three sections.

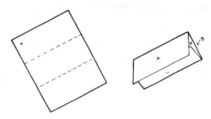

4 Door Book

1. Fold a sheet of paper into a Shutter Fold. See page 1.

2. Fold it into a Hamburger.

3. Open the Hamburger and cut the Valley folds on the Shutters only, creating four tabs.

4. Refold it into a Hamburger, with the fold at the top. Decorate the top sheet as the cover.

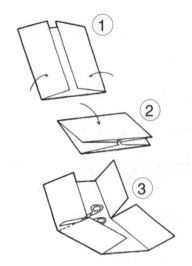

Bound Book

1. Take two sheets of paper and fold them like a Hamburger.

2. Place the folds side-by-side allowing 1/16" between the mountain tops. Mark both folds 1" from the outer edges.

3. On one of the folded sheets "cut up" from the top and bottom edges to the marked spots on both sides.

4. On the second folded sheet, start at one of the marked spots and "cut out" the fold between the two marks. Do not cut into the fold too deeply, only shave it off.

5. Take the "cut up" sheet and roll it. Place it through the "cut out" sheet and then open it up. Fold the bound pages in half to form a book.

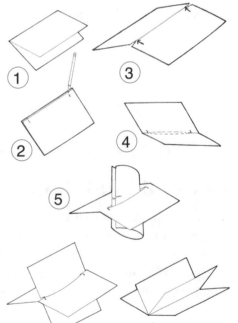

Variation...

To make a larger book use additional sheets of paper, marking each sheet as explained in #3.
Use equal number of sheets for the "cut up" and "cut out."
Place them one on top of the other and follow directions 4 through 5.

Side-by-Side

Some books can easily grow into larger books by gluing them side-by-side. Make two or more of these books. Be sure the books are closed, then glue the back cover of one book to the front cover of the next book. Continue in this manner, making the book as large as needed. Glue a cover over the whole book.

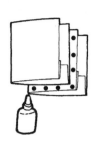

The *Lots of Science Library Book* Shelf

Make a bookshelf for the *Lots of Science Library Books* using an appropriate sized box or following the instructions below.

1. Begin with an 11" x 12" piece of posterboard or cardboard.
   Mark lines 3" from the edge of each side. Fold up on each folded line.
   Cut on the dotted lines as indicated in illustration 1.

2. Refold on the line.
   Glue the tabs under the top and bottom sections of the shelf. See illustration 2.
   Cover your shelf with attractive paper.

3. If you are photocopying your *Lots of Science Library Books*, consider using green paper for the covers and the same green paper to cover your bookshelf.

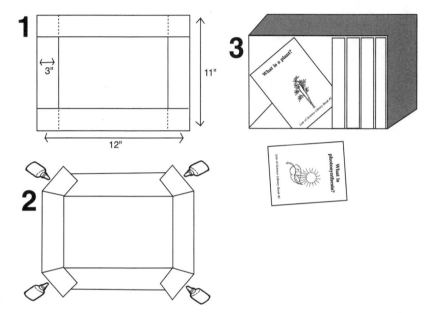

Teacher's Section

Website addresses used as resources in this book are accurate and relevant at the time of publication. Due to the changing nature of the Internet, we encourage teachers to preview the websites prior to assigning them to students.

The authors and the publisher have made every reasonable effort to ensure that the experiments and activities in this book are safe when performed according to the book's instructions. We further recommend that students undertake these activities and experiments under the supervision of a teacher, parent, and / or guardian.

A GREAT LIFE SCIENCE STUDY

Plants Concept Map
Lessons 1-5

Numbers Refer To Lesson Numbers

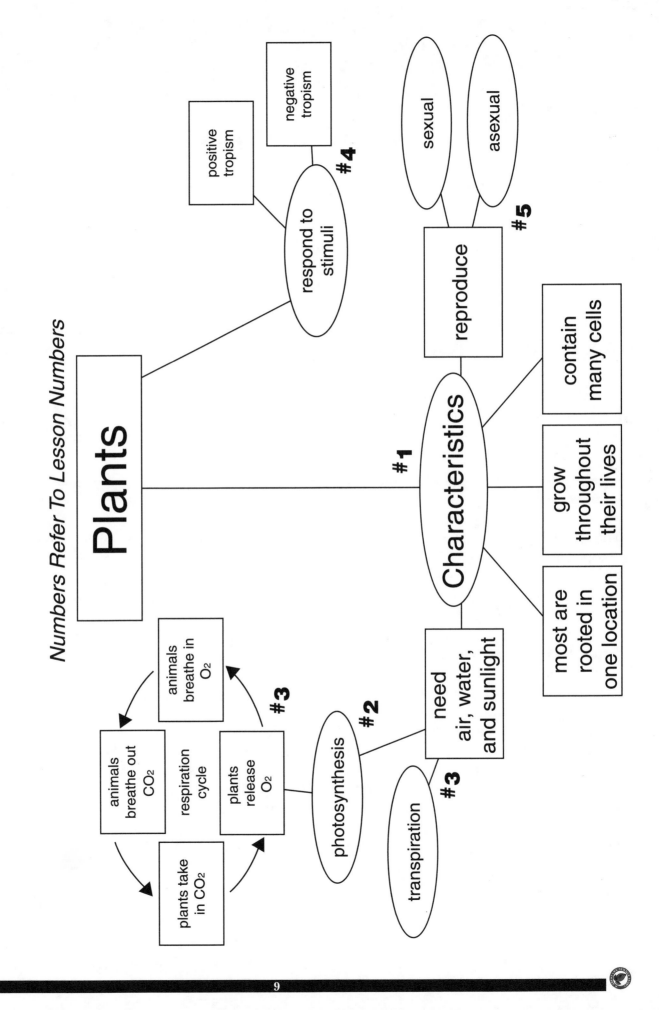

What is a plant?

Plant Concepts:

- All plants need water, sunlight, and air to live and grow.
- All plants grow throughout their lives.
- All plants reproduce plants of their own kind.
- All plants are eukaryotes, or contain more than one cell.
- Most plants are rooted in one location their entire lives unless someone or something moves them.

Vocabulary Words: rooted   air   sunshine   live   plant   cell
grow   *reproduce   *biome (BY ome)

http://images.
botany.org

Read: *Lots of Science Library Book #1.*

Activities:

Investigative Loop – Watch a Plant Grow

Focus Skill: quantitative observations

Lab Materials: 2-3 bean seeds   paper towels   ruler   clear jar filled with soil

Paper Handouts: 2 sheets of 8.5"x11" paper   a 12"x18" sheet of construction paper

Graphic Organizer: Make two Small Question and Answer Books out of 8.5"x11" sheets of paper. Make a Half Book out of a 12"x18" sheet of paper. Glue the Small Question and Answer Books inside the Half Book.

Concepts: Plants use water, soil, air and light to grow. Plant growth can be measured.

Procedure: Sprout seeds between wet paper towels. Plant sprouts around the inside edges of the jar. Keep the soil moist and warm. Place the jar to receive indirect sunlight.

Observations: Choose one seedling to observe and measure as it grows for eight days. Mark your Lab Log for eight days.

Record the Data: Label the front tabs of the Small Question and Answer Books with Day 1: date, Day 2: date, etc. at the bottom. Each day, sketch the plant on the outside tab and label the parts. Under the tab, record the plant growth in standard measurements.

Conclusions: Draw conclusions about the needs of a plant and the progress of its growth.

Communicate the Conclusions: On the back of the Half Book, write a narrative of the life of the bean plant.

Spark Questions: Discuss questions sparked by this lab.

New Loop: Choose one question to investigate further through research, observations or Investigative Loop.

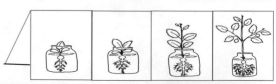

Plants...

Focus Skill: obtaining information

Paper Handouts: 8.5"x11" sheet of paper

a copy of Graphics Page 1

Graphic Organizer: Make a Half Book.

✎ Write/copy *Plants* on the cover. Inside, draw/glue the pictures from Graphics Page 1. Write/dictate clue words about each picture. Examples: *sunlight, air, water, many cells, young plants, rooted.* Orally explain each picture using a complete sentence.

✎✎ Begin the same as ✎ students. Write two sentences that describe the common characteristics of plants.

✎✎✎ Create a title and cover for the Half Book. Inside, summarize information about the common characteristics of plants.

Materials: Nature Guide Book

Paper Handouts: 8.5"x 11" sheet of white paper

Graphic Organizer: Make a Large Question and Answer Book. See page 2 for instructions. This book is used in this and future lessons. As new Large Question and Answer Books are made in future lessons they will be glued "side-by-side" to this book.

Procedure: Go on a nature walk. Guide students in plant observation and aid in identification. Students select two plants to feature in their Plant ID Books. Draw one plant on each tab. Record the plant observations under the tabs.

✎ Copy the name of the plant and draw a picture of the location where it was found.

✎✎ Copy the name of the plant, using the scientific name if possible. Record the following data: date, weather conditions, and location where it was found.

✎✎✎ Complete ✎✎. Hypothesize as to why each was found growing in this location.

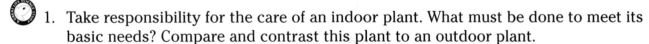

Experiences, Investigations, and Research

Select one or more of the following activities for individual or group enrichment projects. Allow your students to determine the format in which they would like to report, share, or graphically present what they have discovered. This should be a creative investigation that utilizes your students' strengths.

1. Take responsibility for the care of an indoor plant. What must be done to meet its basic needs? Compare and contrast this plant to an outdoor plant.

2. Research an unusual plant. Suggestions: Venus flytrap, butterwort, pitcher plant, or sundew. Share this information with three people in a 24 hour period.

3. Most states and countries use flowers and trees as symbols. Investigate the symbols that represent your area and a location you would like to visit.

4. The olive branch is a symbol for peace; find out why.

What is photosynthesis?

Plant Concepts:

- All plants contain chlorophyll. Look at a green plant as you review these concepts.
- Chlorophyll gives the plant its green color.
- Most photosynthesis takes place in the leaves of the plant.
- Chlorophyll absorbs sunlight energy.
- The sunlight energy breaks down water into its two elements, hydrogen and oxygen.
- The plant releases oxygen into the air.
- The plant absorbs carbon dioxide from the air and mixes it with hydrogen to make glucose, or sugar.
- In addition to sugar, the plant makes other food substances that it either uses or stores for future use.
- The stored food is called sap.

Teacher's Note: An alternative assessment suggestion for this lesson is found on pages 78-79. If Graphic Pages are being consumed, photocopy assessment graphics needed first.

Vocabulary Words: green   plant   sugar   sap   absorbs   hydrogen
oxygen   *carbon dioxide   *chlorophyll   *photosynthesis (foh toh SIN theh sis)

Read: *Lots of Science Library Book #2.*

Activities:

Photosynthesis – Graphic Organizer

Focus Skill: sequencing a process

Paper Handouts: 8.5"x11" sheet of paper   a copy of Graphics 2A – D

Graphic Organizer: Make a Small Question and Answer Book. Draw/glue the pictures in the correct order on the front tabs. Under the tabs, write/dictate clue words about the process of photosynthesis.
1. Plant leaves are green because of chlorophyll. *(plant, green)*
2. Plants need sunlight and water. *(sunlight, water)*
3. Plants use sunlight and water to make their food. *(make food)*
4. Plants move the food they make throughout their parts. *(move and use food)*

Paper Handouts: two 8.5"x11" sheets of paper   a copy of Graphics 2A – H
a 12"x18" sheet of construction paper

Graphic Organizer: Make two Small Question and Answer Books. Draw/glue the pictures in the correct order to illustrate the process of photosynthesis. Make a Half Book from the 12"x18" paper. Glue the small Question and Answer Books inside the Half Book.

✎✎ Explain the process of photosynthesis by writing clue words or phrases under each tab. Orally explain the process using complete sentences.

✎✎✎ Research photosynthesis. Under each tab explain that step in the process. Explain how photosynthesis changes from day to night.

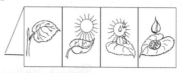 

Investigative Loop – Chlorophyll in Plants – Lab 2-1

Focus Skill: drawing conclusions from observations

Lab Materials: a clear glass half full of rubbing alcohol   a fresh leaf

Paper Handouts: 8.5" x 11" sheet of paper   a copy of Lab Graphic 2-1
    Lab Record Cards (index cards or 1/4 sheets of paper)

Graphic Organizer: Make the Pocket Book. See page 2 for instructions. This is the student's Lab Book. In future Lessons, Pocket Books will be made and glued side-by-side to this one. Glue Lab Graphic 2-1 on the left pocket.

Concept: The chlorophyll that makes plants green can be extracted for observation.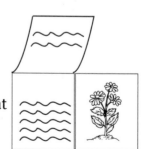

Prediction: If chlorophyll is extracted from a leaf into alcohol, what color will it be?

Procedure: Place a freshly picked leaf in the glass of alcohol. (Note: Dip the leaf in boiling water before placing it in the alcohol to speed up the process.) Set a timer and check the leaf every hour for several hours. Check it the next day.

Observations: Observe the leaf before, during, and after placing it in the alcohol.
    How does the leaf change? How does the alcohol change?

Record the Data: On your Lab Record Cards, write Lab 2-1, the date, and leaf observations.
    Diagram the leaf and alcohol as they were before and after the experiment.

Conclusions: Draw conclusions from your observations.

Communicate the Conclusions: On a Lab Record Card, explain how your observations led to the conclusions. Place the Lab Record Cards in the Lab Book for Lab 2-1.

Spark Questions: Discuss questions sparked by this lab.

New Loop: Choose one question to investigate further,
    Or repeat the above procedure using leaves of different plants in new glasses of alcohol. Compare the color of the alcohol for each plant. Make a Lab Record Card for each plant.

Add to your Plant ID Book

Materials: Nature Guide Book

Paper Handouts: 8.5"x11" sheet of white paper   student's Plant ID Book

Graphic Organizer: Make a Large Question and Answer Book. Glue it side-by-side to the Plant ID Book made in the previous lesson.
    Students select two plants to feature in their Plant ID Books. Draw one plant on each tab. Record the plant observations under the tabs. See Lesson 1 Activities section.

Experiences, Investigations, and Research

Select one or more of the following activities for individual or group enrichment projects. Allow your students to determine the format in which they would like to report, share, or graphically present what they have discovered. This should be a creative investigation that utilizes your students' strengths.

 1. Write a word with plants. Find a good location for plants to grow, write a word in the dirt, and sprinkle carrot seeds in the small furrows. Cover the seeds with dirt and water. Check your word growth daily.

 2. Chloroplasts are the tiny parts of the plant cell that contain chlorophyll. Research the structure of a plant cell. Focus on chloroplasts. Make a Half Book for this project. Draw a plant cell on the cover. Inside, illustrate the cell and label the parts. Describe each part and its function in the cell.

3. Read *The Tale of Peter Rabbit* by Beatrix Potter.

4. Begin reading *My Side of the Mountain* by Jean Craighead George or *Swiss Family Robinson* by Johann David Wyss.

http://sierra.com
/sierrahome/
gardening/

Notes

What are respiration and transpiration in plants?

Plant Concepts:

- The respiration cycle is the process of breaking down food molecules into gases, water, and energy.
- Plants take in carbon dioxide and give off oxygen in this process.
- Since all animals inhale oxygen and exhale carbon dioxide, the respiration cycle of plants is essential to all other life.
- During the transpiration cycle, water is released through pores on the underside of leaves.

Vocabulary Words: inhale   exhale   oxygen   *carbon dioxide   *respiration
*transpiration   *stomata (STO muh tuh)

Read: *Lots of Science Library Book #3.*

Activities:

The Respiration Cycle - Graphic Organizer

Focus Skill: cause and effect
Paper Handouts: 4.25"x11" sheet of paper   a copy of Graphic 3A
Graphic Organizer: Make a Shutter Fold Book. Cut Graphic 3A in half and glue one
    half on each shutter, as shown.

✎   Write/dictate clue words about the respiration cycle.
    Example: under the leaf shutter - *gives off oxygen, takes in carbon dioxide*
    under the girl shutter – *breathes in oxygen, breathes out carbon dioxide*
    Orally explain the process using complete sentences.

✎✎   Complete ✎. Under the clue words, write a sentence explaining each part of the
    respiration cycle. On the inside middle section, draw the respiration cycle of an
    animal in its native habitat.

✎✎✎   On the back of each shutter, explain the steps of the respiration cycle. On the inside
    middle section, compare and contrast respiration and photosynthesis. On the back,
    explain why the tropical rain forests are called the "lungs of the earth." Defend your
    statements with facts about the rain forests.

Investigative Loop – Leaves Give Off Oxygen – Lab 3-1

Focus Skill: diagramming
Lab Materials: clear glass   newly cut leaf   magnifying glass
Paper Handouts: Lab Book   new Lab Record Cards   a copy of Lab Graphic 3-1

Graphic Organizer: Glue Lab Graphic 3-1 on the right pocket of the Lab Book.
Concept: Leaves release oxygen.
Research: Read the *Lots of Science Library Book #3*. Review how leaves release oxygen.
Predictions: Predict where you will find oxygen being released from the leaf.
Procedure: Put the leaf in a clear glass of water. Place the glass in indirect sunlight.
    In four hours, examine the leaf with a magnifying glass.

Observations: Describe the leaf using qualitative observations before, during, and after the lab.
    See *How to Use This Program, Investigative Loop* for more information.
Record the Data: On a Lab Record Card, record Lab 3-1, the date, and observations. Sketch the
    leaf and describe it qualitatively.
Conclusions: Draw conclusions about what you observed in the leaf based on your knowledge of
    photosynthesis and respiration. **Possible answer: The oxygen bubbles on the leaf and in the water**
    were released by the leaf.
Communicate the Conclusions: On another Lab Record Card, write your conclusions. Include a
    sketch of the leaf after one hour in the water. Put the Lab Record Cards in the
    Lab Pocket for Lab 3-1.
Spark Questions: Discuss questions sparked by this lab.
New Loop: Choose one question to investigate further.
    Or complete the procedure above except put the glass in a dark location. After the
    observations and conclusions are made, compare and contrast them with the previous lab.
    Note: There should be fewer or no bubbles as the lack of sunshine slows down the photosynthesis
    process thus slowing down the release of oxygen.

Investigative Loop – Leaves Transpire - Lab 3-2

Focus Skill: questioning
Lab Materials: a plastic reclosable bag   newly cut leaf   magnifying glass
Paper Handouts: 8.5"x11" sheet of paper   a copy of Lab Graphic 3-2
    Lab Book   new Lab Record Cards
Graphic Organizer: Make a Pocket Book and glue it side-by-side to the students' Lab Book.
    Glue Lab Graphic 3-2 to the left pocket of the new Pocket Book.
Question: How do leaves release water?
Research: Review the *Lots of Science Library Book #3* and review the question.
Predictions: Predict if and where water from this leaf will be released. Write your prediction on a
    Lab Record Card labeled Lab 3-2.
Procedure: Place a newly cut leaf in a plastic reclosable bag. Examine the leaf with a magnifying
    glass in three hours.
Observations: Make qualitative observations of the leaf. See *How to Use This Program,*
    Investigative Loop for more information.
Record the Data: On a Lab Record Card, record Lab 3-2 and the date. Enter the time and a brief
    description of each observation. Draw the leaf at the final observation time.
Conclusions: Explain why the leaf looked as it did after three hours. Draw conclusions about the
    release of water by leaves. **Note: The moisture on the underside of the leaf indicates it is**
    releasing water through transpiration.
Communicate the Conclusions: On a Lab Record Card, compare your observations and
    conclusions with your predictions. Share your Lab Record Cards with one person who did
    not participate in the Lab. Store them in the Lab Book for Lab 3-2.
Spark Questions: Discuss questions sparked by this lab.
New Loop: Choose one question and investigate it further.

Choose a Topic: Select two or all four of the following biomes: tundra, grasslands, rain forest, desert.

Conduct research, document your findings, and report on the following:
Average annual temperature, average annual precipitation, effect of geographic locations, common native plants, and interesting facts about the vegetation in general.

Paper Handouts: One sheet of 12" x 18" construction paper for each biome researched a copy of these Graphics for each biome: world map, rain gauge, thermometer, plants for that biome, index cards, a copy of the world map in the Graphics Pages.

Graphic Organizer: Make a Pyramid Project diorama for each biome researched. Record annual temperatures and rainfall on the graphics, and color the small world maps to indicate geographic locations of each biome. Glue these inside the biome dioramas. Select appropriate plants to glue in the foreground. Glue your pyramid dioramas together to make a 2 or 4 part display. Write additional information on index cards to display on the table around the diorama. Compare and contrast the environments. Share what you have learned with others.

Experiences, Investigations, and Research

Select one or more of the following activities for individual or group enrichment projects. Allow your students to determine the format in which they would like to report, share, or graphically present what they have discovered. This should be a creative investigation that utilizes your students' strengths.

1. Design an Investigative Loop to answer this question: How much water is lost by leaves in 24 hours? Example: After sunset, tie a clear plastic bag over a cluster of leaves on a living shrub. At the end of 24 hours, remove the bag and describe the amount of water.

2. Investigate the flora and fauna of the tropical rain forest and how they affect one another.

3. Discover which rain forest plants are used for commercial products.

4. Discover which rain forest plants are used for medical purposes.

5. Predict what would happen if the rain forests were cut down.

6. Read *The Great Kapok Tree: A Tale of the Amazon Rain Forest* by Lynne Cherry.

Notes

How do plants respond to the world around them?

Plant Concepts:

- Plants do not have sensory organs like animals.
- Plants do respond to certain stimuli in their environment.
- When plants respond to stimuli it is called tropism.
- When a plant grows toward light it is called phototropism.
- When a plant grows toward water it is called hydrotropism.

Vocabulary Words: sensory   special   respond   *mobility   *stimuli
*tropism (TRO piz um)   *phototropism   *hydrotropism

Read: *Lots of Science Library Book #4.*

Activities:

Tropism - Experiment

Ask your students to review the needs of a plant and think of two stimuli that a plant will respond to with tropism. Assist your students in designing an Investigative Loop for each stimulus or complete the ones listed below.

Investigative Loop – Tropism and Sunshine – Lab 4-1

Focus Skill: cause and effect
Lab Materials: a growing plant   a cardboard box
Paper Handouts: Lab Book   new Lab Record Cards
                 a copy of Lab Graphic 4-1   Lab Log
Graphic Organizer: Glue Lab Graphic 4-1 on the right pocket in the Lab Book.
Question: How does a plant respond to sunlight?
Research: Read the *Lots of Science Library Book #4* and review this question.
Predictions: Predict how a plant will respond to sunlight. Write your prediction on a Lab Record Card labeled Lab 4-1.
Procedure: Cut a 2" square hole in the side of the box. Place the plant inside the box with the lid on top. Place the box so that sunlight comes in the hole. Water the plant daily for two weeks.
Observations: Each day, check the plant and note any changes. Mark your Lab Log for two weeks.
Record the Data: Each day, on a Lab Record Card write Lab 4-1, the date, and your observations of the plant.

Conclusions: Compare the plant at the end of the lab to how it looked two weeks ago. What caused the plant to grow as it did? What effect does sunlight have on a plant's growth? Compare the plant to your predictions.

Communicate the Conclusions: On a Lab Record Card, write your conclusions about plants and sunlight. Include applications for these conclusions. Store the Lab Record Cards in the Lab Book for Lab 4-1.

Spark Questions: Discuss questions sparked by this lab.

New Loop: Choose one sparked question to investigate further.

Investigative Loop – Water and Tropism – Lab 4-2

Focus Skill: cause and effect

Lab Materials: a growing plant   a clear glass jar filled with soil   a side of a milk carton

Paper Handouts: Lab Book   Lab Record Cards   Lab Log   a copy of Lab Graphic 4-2

Graphic Organizer: Make another Pocket Book and glue it side-by-side to the Lab Book. Glue Lab Graphic 4-2 on the left pocket of the new Pocket Book.

Question: How does a plant respond to water?

Research: Read the *Lots of Science Library Book #4* and review the question.

Predictions: Predict how a plant will respond to water. Write your prediction on a Lab Record Card that is labeled Lab 4-2.

Procedure: Cut a piece of a milk carton to fit inside of the jar. Punch five holes in the milk carton piece, close to one side. Slide the milk carton piece in the middle of the soil, with the punched holes to the front of the jar. Place the growing plant on one side of the milk carton divider near the front of the jar. Do not water that side of the jar. Keep the other side of the jar moist. See illustration below.

Observations: Each day, check the plant and note any changes. Mark your Lab Log for two weeks.

Record the Data: Each day, on a Lab Record Card write Lab 4-2, the date, and your observations of the plant. Sketch illustrations of the plant if desired.

Conclusions: Compare the plant at the end of the lab to how it looked two weeks ago. What caused the plant to grow as it did? What effect does water have on a plant's growth? Compare the plant to your predictions.

Communicate the Conclusions: On a Lab Record Card, write your conclusions about plants and water. Include applications for your lab conclusions. Store the Lab Record Cards in the Lab Book for Lab 4-2.

Spark Questions: Discuss questions sparked by this lab.

New Loop: Choose one sparked question to investigate further.

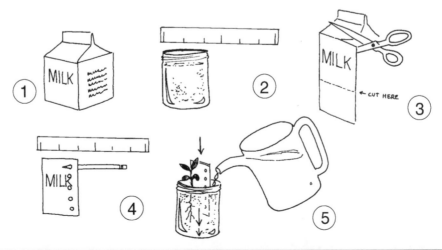

Experiences, Investigations, and Research

Select one or more of the following activities for individual or group enrichment projects. Allow your students to determine the format in which they would like to report, share, or graphically present what they have discovered. This should be a creative investigation that utilizes your students' strengths.

 1. Germinate a seed. When it begins to grow, tape a quarter to the top of it. Do you think it will continue to grow or stop growing?

 2. Investigate how these plants respond to the length of day and night: larkspur, chrysanthemum, and snapdragon.

 3. Observe how sunflowers follow the sun.

 4. Observe how bushes and trees compete for light. Notice how bushes have the most green leaves in areas where sunlight reaches them.

 5. Research negative tropism. In a 3 Tab Book, compare and contrast positive and negative tropism in plants. Use examples and illustrations.

Notes

A GREAT LIFE SCIENCE STUDY

How do plants reproduce?

Plant Concepts:

- Reproduction is the process of producing new forms of a living thing.
- All plants reproduce sexually. Some plants also reproduce asexually.
- Sexual reproduction involves fusing a male sex cell, or gamete, and a female sex cell, or gamete.
- Plants reproduce sexually with seeds or spores.
- Plants that reproduce asexually use their stems, roots, or leaves to create new plants.

Teacher's Note: An alternate assessment suggestion for this lesson is found on pages 78-79. If Graphic Pages are being consumed, photocopy assessment graphics needed first.

Vocabulary Words: reproduce   *gamete   *sexual reproduction
*asexual reproduction   *rhizomes (RY zomes)   *zygote (ZY gote)

Read: *Lots of Science Library Book #5.*

Activities:

How Plants Reproduce - Graphic Organizer

Focus Skill: sorting by characteristics
Paper Handouts: 12"x18" sheet of construction paper   half sheets of notebook paper
✎ a copy of Graphics 5B, C ✎✎ ✎✎✎ a copy of Graphics 5A, B
Graphic Organizer: Make a Large Pocket Book. Title it *Reproduction Plant Book.*

✎ On the left pocket glue Graphic 5C. On the right pocket glue Graphic 5B. On notebook paper write/dictate clue words for plants that reproduce through seeds and spores. Include examples of plants that reproduce in that manner. Do the same for plants that reproduce through their stems, roots, and leaves.

✎✎ On the left pocket glue Graphic 5A and label it *sexual reproduction*. On the right pocket glue Graphic 5B and label it *asexual reproduction*. On notebook paper write clue words and sentences explaining each type of reproduction.

✎✎✎ Complete ✎✎. Research these plants: sweet potato, iris, raspberry, kalanchoe, daffodil, and gladiolus. On the half sheets of notebook paper describe how each reproduces. Use illustrations if possible. Place each paper in the correct pocket of the Reproduction Plant Book.

Investigative Loop – Asexual Reproduction – Lab 5-1

Focus Skills: compare and contrast, cause and effect
Lab Materials: a plant cutting that will root (such as an African violet leaf, geranium leaf, ivy stem, or carrot top)
a plant cutting that will not root (such as a flower on a stem)
two clear glasses of water

Paper Handouts: Lab Book   Lab Record Cards   a copy of Lab Graphic 5-1   Lab Log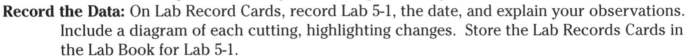

Graphic Organizer: Glue Lab Graphic 5-1 on the right pocket in the Lab Book.

Question: Do plant cuttings respond in the same manner to water?

Research: Read the *Lots of Science Library Book #5* and review asexual reproduction.

Predictions: On a Lab Record Card, predict how you think each plant will respond to being cut and put in water. Diagram your predictions if desired.

Procedure: Put each cutting into a glass of water.

Observations: Each day, observe the plant cuttings. Look for similarities and differences in them. Observe how they are the same as they were at the beginning of this lab. How do they differ? Mark your Lab Log for 10 days.

Record the Data: On Lab Record Cards, record Lab 5-1, the date, and explain your observations. Include a diagram of each cutting, highlighting changes. Store the Lab Records Cards in the Lab Book for Lab 5-1.

Conclusions: After 2 weeks of recording the data, compare your Lab Record Cards. What conclusions can you draw from your observations? How are the plant cuttings the same? How do they differ? How do you explain this difference? How does the outcome compare with your predictions?

Communicate the Conclusions: Pretend you are a newspaper reporter with the late breaking news about these plant cuttings. In a Half Book, write your article explaining this Investigation Loop. Include the observations and conclusions made. Compare the two cuttings, using illustrations if desired. Fold the Half Book and store it in the Pocket for Lab 5-1.

Spark Questions: Discuss questions sparked by this lab.

New Loop: Choose one sparked question to investigate further, or complete the above procedure with two or more different plant cuttings.

Seed Collection

Begin a seed collection to be used in Lesson 13. Collect seeds from plants that can be identified. Label each seed by the plant name. Safely store them in separate plastic bags.

Add to your Plant ID Book

Materials: Nature Guide Book

Paper Handouts: 8.5"x11" sheet of white paper   Plant ID Book

Graphic Organizer: Make a Large Question and Answer Book. Glue it side-by-side to the Plant ID Book. Students select two plants to feature in their Plant ID Books. Draw one plant on each tab. Record the plant observations under the tabs including information about how each plant reproduces.

Experiences, Investigations, and Research

Select one or more of the following activities for individual or group enrichment projects. Allow your students to determine the format in which they would like to report, share, or graphically present what they have discovered. This should be a creative investigation that utilizes your students' strengths.

 1. List the possible negative aspects of asexual reproduction.

 2. Investigate plant grafting. How is it done and why is it useful commercially?

 3. Fruits have seeds. Have you ever seen banana seeds? Research how bananas reproduce.

Notes

Plants Concept Map
Lessons 6-11

Numbers Refer to Lesson Numbers

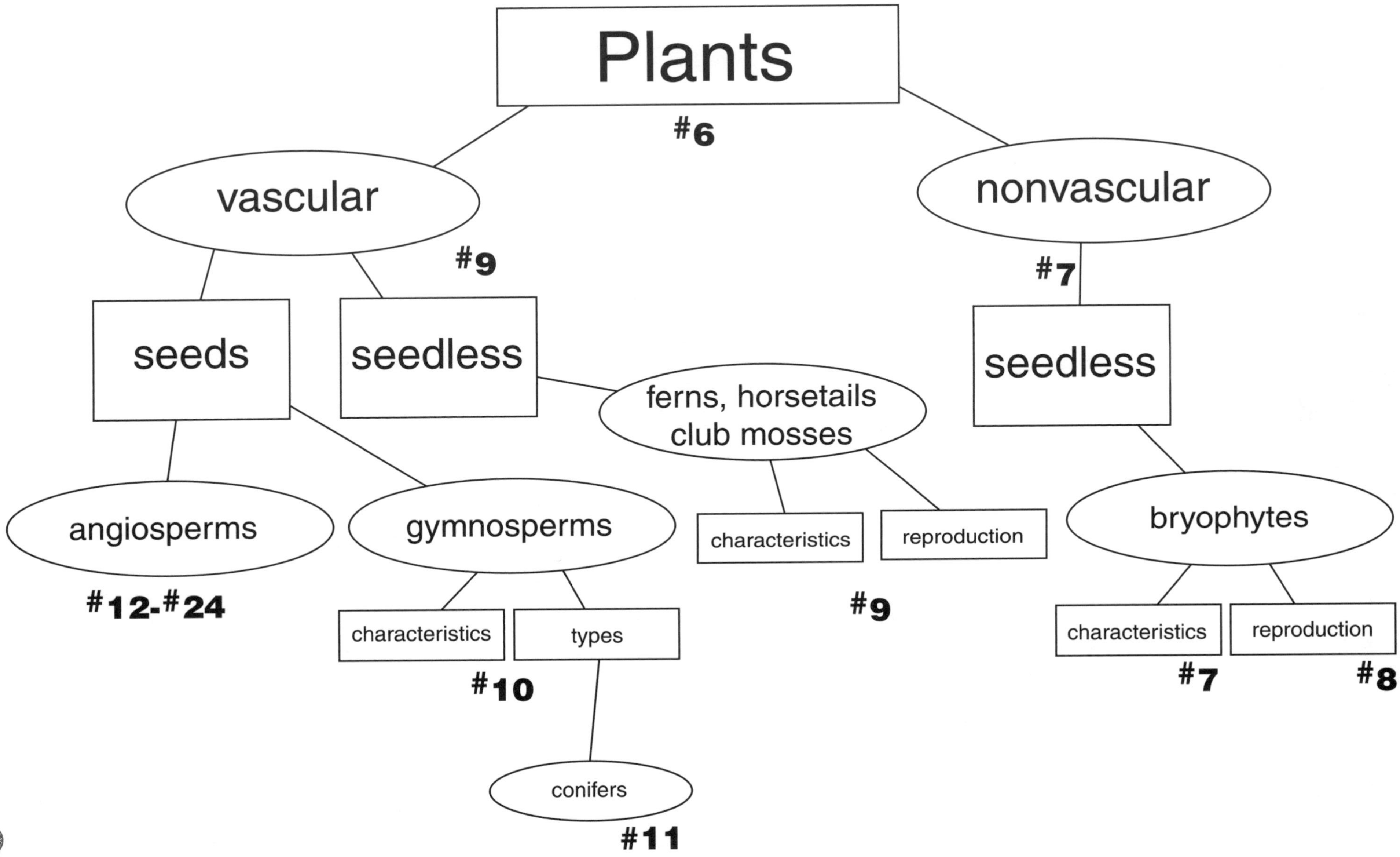

Plants

#6

vascular

#9

- seedless
  - gymnosperms
    - characteristics **#10**
    - types
      - conifers **#11**
  - ferns, horsetails club mosses
    - characteristics
    - reproduction

    #9

- seeds
  - angiosperms **#12-#24**

nonvascular

- seedless **#7**
  - bryophytes
    - characteristics **#7**
    - reproduction **#8**

What are the different types of plants?

Plant Concepts:

- Plants are classified by similarities in structure and reproduction.
- Plants are divided into four main groups:
  - Bryophytes
  - ferns, horsetails, and club mosses
  - nonflowering seed plants (Gymnosperms)
  - flowering seed plants (Angiosperms)

Bryophytes	

Vocabulary Words: moss   fern   flower   seed   classify

*botanist   *gymnosperm (JIM nuh spurm)   *angiosperm (AN jee uh spurm)

Ferns Horsetails and Club Mosses	

Read: *Lots of Science Library Book #6.*

Activities:

Types of Plants - Graphic Organizer

Paper Handouts: 4 sheets of 12"x18" construction paper

Graphic Organizer: Fold all of the pieces of paper into Hamburgers. Turn each paper so the fold is on the left side. Each book is for a group of plants. Ask your students to open each book and write or copy the name of the group on the top of the left page.

Seed Plants Gymnosperms	

1. *Bryophytes*
2. *Ferns, horsetails, and club mosses*
3. *Seed Plants Gymnosperms*
4. *Flowering Seed Plants Angiosperms*

Flowering Seed Plants Angiosperms	

Teacher's Note: These books will be made into a Large Graphic Organizer Project that will grow in the next 14 lessons. At that time your students will have information on the four major plant groups.

Activity Materials: an assortment of objects to classify or sort using common characteristics; such as wood objects, metal objects, and plastic objects or make a copy of Graphics Page 6.

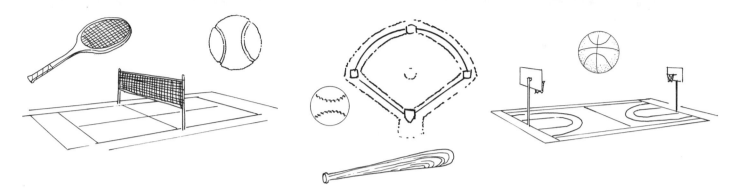

Activity: Discuss the objects/pictures. Use qualitative observations in this activity. Discuss how they look, what material they are made out of, what they are used for, and any other characteristic that fits the assortment.

Ask your students to choose a way to classify all the objects. Remember, the characteristics that are chosen will determine what object fits into a group.

When this is completed, ask your students to group the objects using different physical characteristics or qualities.

Sort one of the following: package of mixed beans, mixed candy, granola cereal.

Experiences, Investigations, and Research

Select one or more of the following activities for individual or group enrichment projects. Allow your students to determine the format in which they would like to report, share, or graphically present what they have discovered. This should be a creative investigation that utilizes your students' strengths.

1. There are about 300,000 species of plants in the world. Research how many species there are of each plant group. Make a pie graph of the information.

2. Investigate aquatic plants. Compare and contrast them to land plants. To observe aquatic plants, go to a pet or aquarium store.

3. Determine what characteristics were used by scientists to sort plants into four groups.

4. Read *Jack and the Beanstalk.*✎ ✎✎

What are bryophytes?

Plant Concepts:

- Mosses and liverworts are bryophytes, meaning, "moss-like" plant.
- Bryophytes do not have true stems, leaves, and roots.
- Bryophytes are simple plants that grow low to the ground in moist areas.
- Bryophytes absorb water and nutrients through their cell walls. They are nonvascular plants.
- Bryophytes are anchored to the ground by rhizoids instead of roots.

Vocabulary Words: moss   simple   moist   cells   anchored
*bryophytes (BRY ah fites)   *rhizoid (RY zoid)

Read: *Lots of Science Library Book #7.*

Activities:

Investigative Loop – How do Bryophytes Absorb Water? Lab 7-1

Focus Skill: discovering a process
Lab Materials: paper towels   food coloring   eyedropper
Paper Handouts: 8.5"x11" sheet of paper   Lab Book   Lab Record Cards
                a copy of Lab Graphic 7-1   a copy of Graphic 7A
Graphic Organizer: Make another Pocket Book and glue it side-by-side to the
        Lab Book. Glue Lab Graphic 7-1 on the left pocket of the new book.
Concept: Bryophytes absorb and transport water through their cell walls.
        They are nonvascular plants.
Research: Read *Lots of Science Library Book #7.* Review how bryophytes
        absorb and transport water and nutrients.
Procedure: Cut paper towels into 20 of these shapes to represent moss plants. See picture. Place
        the paper towel pieces on a table or plate. Using the eyedropper, put several drops of
        colored water on the paper towel pieces and near the paper towel pieces.
Observations: Examine the paper towel pieces closely. Discover how the water moves through the
        pieces, how the water moves from one piece to another, and how far the water travels in
        the pieces.
Record the Data: Label a Lab Record Card 7-1, the date, describe your observations, and diagram
        the procedure.
Conclusion: If mosses absorb water similarly to the paper towel pieces, why do they have to live
        in moist places? Why do they have to live low to the ground? Why are mosses small
        plants?
Communicate the Conclusions: On another Lab Record Card, explain how bryophytes absorb
        water and nutrients based on your observations. Draw conclusions about their
        environment and characteristics based on this lab.

Spark Questions: List questions sparked by this lab.
New Loop: Choose one question to investigate further with research, observations, or an experiment.

Glue the 'cellular plant' graphic (7A) on the top of the left page in the Bryophyte Book. Label the graphic *nonvascular plant*.

What is a Bryophyte? - Graphic Organizer

Focus Skill: describing the characteristics of a group
Paper Handouts: 8.5"x11" sheet of paper   a copy of Graphics 7B, C, D,
            Bryophyte Book made in Lesson 6
Graphic Organizer: Make a Half Book. Glue Graphic 7B on the cover and title the book. Inside, glue Graphics 7C and D on the top section.

✐   Inside the book, glue the paper towel mosses from the previous Investigative Loop. Draw rain saturating the moss. Orally discuss the characteristics of moss.

✐✐   Label the parts of the moss. Write clue words that describe bryophytes.
Examples: *simple plants, absorb through cells, anchored, low to ground, moist places.* Use the clue words to write sentences describing bryophytes.

✐✐✐   Label the parts of the moss. Write sentences describing bryophytes. Investigate liverworts as well as mosses. Compare and contrast the two types of plants.

Glue this Half Book onto the bottom of the left page in the Bryophyte Book.

Finding Mosses - Plant ID Book

Materials: Nature Guide Book
Paper Handouts: 8.5"x11" sheet of white paper   Student's Plant ID Book
Graphic Organizer: Make a Large Question and Answer Book. Glue it side-by-side to the student's Plant ID Book.
Students select one bryophyte to feature in their Plant ID Books. Draw the plant on the left tab. Record your plant observations under the tab. The right tab will be used in Lesson 8.

Experiences, Investigations, and Research

Select one or more of the following activities for individual or group enrichment projects. Allow your students to determine the format in which they would like to report, share, or graphically present what they have discovered. This should be a creative investigation that utilizes your students' strengths.

1. Place peat moss in a bucket or container. Add 1/2 cup of water at a time until the peat moss is saturated. How much water did it hold?

2. Explain why mosses are used in floral arrangements and gardens.

3. Outline how a lava bed could become a rain forest over time. Explain the importance of pioneer plants, like mosses, in this process. At what point would animals become a part of the environment?

4. Investigate why peat moss bogs are found mainly in the Northern Hemisphere.

5. Research peat bog mummies and explain why they are so well preserved.

How do bryophytes reproduce?

Plant Concepts:

- Bryophytes reproduce through spores and gametes, or sex cells.
- Water is needed for male gametes to fertilize female gametes.
- The fertilized embryo becomes a new plant, or sporophyte, that remains attached to the parent plant.
- Sporophytes produce spores in a spore case, or capsule, at the top of their stalks.
- When mature, spore cases open and spores are released.
- Spores germinate and grow into new bryophytes.

Vocabulary Words: spore   reproduce   water   spore case

*sporophyte (SPO roh fite)

Read: *Lots of Science Library Book #8.*

Activities:

How do Bryophytes Reproduce? - Graphic Organizer

Focus Skill: sequencing a process
Paper Handouts: 3 sheets of 8.5"x11" paper   a copy of Graphics 8A – G
                        Bryophyte Book made in Lesson 6
Graphic Organizer: Make a Layered Look Book using the three papers. Cut the book in half as shown in the diagram. Use half of the book now and half in Lesson 9. On the tabs of the book, draw/glue the steps of reproduction in bryophytes. **Correct order: 8A, 8E, 8C, 8D, 8B, 8F.**

✎   Color the pictures while discussing the process of reproduction in these plants.

✎✎   On each tab, write clue words for each step in the top section.
      Examples: *male and female cells, cells join, plants grow on parent plant, spores are made, spores are released, new plant grows.* Where space allows, write a complete sentence explaining the step in the reproduction process of bryophytes.

✎✎✎ Conduct further research on the reproduction of bryophytes. Write about each step in the process on the tabs. Include the conditions needed for fertilization, how the spores travel, and the likelihood of a spore growing into a new bryophyte.

      Glue the Layered Look Book onto the right page of the Bryophyte Book.
      Glue the 'spores graphic' (8G) to the top of that page.
      Store for future use.

Continue with the Reproduction Pocket Book made in Lesson 5 by asking your students to make a card for bryophytes. Put it in the sexual reproduction pocket of the book.

Finding Spore Cases - Plant ID Book

Materials: Nature Guide Book
Paper Handouts: Plant ID Book
Graphic Organizer: Use the right tab in the Plant ID Book.
    Students select a spore case on a bryophyte plant to feature in their Plant ID Books. Draw it on the right tab. Record your plant observations under the tab.

Experiences, Investigations, and Research

Select one or more of the following activities for individual or group enrichment projects. Allow your students to determine the format in which they would like to report, share, or graphically present what they have discovered. This should be a creative investigation that utilizes your students' strengths.

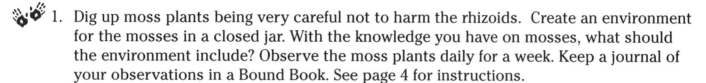 1. Dig up moss plants being very careful not to harm the rhizoids. Create an environment for the mosses in a closed jar. With the knowledge you have on mosses, what should the environment include? Observe the moss plants daily for a week. Keep a journal of your observations in a Bound Book. See page 4 for instructions.

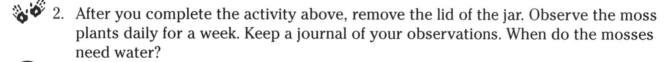 2. After you complete the activity above, remove the lid of the jar. Observe the moss plants daily for a week. Keep a journal of your observations. When do the mosses need water?

3. Compare and contrast the mosses in a closed environment to the mosses in an open environment.

What are ferns and how do they reproduce?

Plant Concepts:

- Ferns are vascular plants with true stems, roots, and leaves.
- Water and nutrients are transported throughout these plants in xylem and phloem vascular tissues.
- Xylem tissues carry water from the roots to the stem and leaves.
- Phloem tissues carry water and glucose from the leaves to the rest of the plant.
- Ferns reproduce with spores and gametes.

Teacher's Note: An alternative assessment suggestion for this lesson is found on pages 78-79. If Graphic Pages are being consumed, photocopy assessment graphics needed first.

Vocabulary Words: fern   water   tissue   stem   root   leaf/leaves   *xylem (ZY lum)   *phloem (FLO em)   *vascular (VAS kuh ler)

Read: *Lots of Science Library Book #9.*

Activities:

What is a Fern? - Graphic Organizer

Focus Skill: explaining characteristics
Paper Handouts: 8.5"x11" sheet of paper   a copy of Graphics 9B, I, and J
                Fern Book made in Lesson 6
Graphic Organizer: Make a Half Book. Glue Graphic 9J on the cover and title it.
        Glue Graphic 9I on the inside of the book.

✎ Inside the book, label the stem, leaves, and roots of the fern. Find a fern and use 2" tape to attach it to the inside of the book.

✎✎ Inside the book, label the parts of the fern. Write clue words about fern plants: *grows tall, vascular system.* Describe where you see ferns growing.

✎✎✎ Inside the book, write about ferns using your vocabulary words. Research horsetails and club mosses. Explain why these are called seedless vascular plants.

Glue the Half Book onto the bottom of the left page in the Fern Book.
Glue the 'vascular graphic' (9B) to the top of that page.
Label the graphic *vascular plant.*

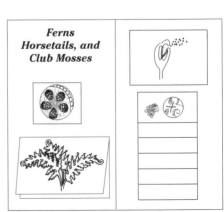

How do Ferns Reproduce? - Graphic Organizer

Focus Skill: sequencing a process
Paper Handouts: the Layered Look Book made in Lesson 8   a copy of Graphics 9A and C – H
Graphic Organizer: On the tabs of the book, draw/glue the steps of reproduction in ferns.
Correct order: 9D, 9H, 9E, 9F, 9G, 9C.

✎ Color the pictures while discussing the process of reproduction.

✎✎ On the tabs, write clue words about each step: *spores under leaf, spores fall and grow, small plant, male and female cells, cells join, new fern grows.* Where space allows, write a complete sentence for the step using the clue words.

✎✎✎ Conduct further research about the reproduction of ferns. Write one or two sentences on each tab about that step in the process. Include the conditions needed for fertilization, how the spores travel, and the likelihood of a new fern growing to maturity.

Glue the Layered Look Book on the right page of the Fern Book.
Glue the 'spores graphic' (9A) to the top of that page. Store for future use.

Investigative Loop – Vascular Plants at Work– Lab 9-1

Focus Skill: compare and contrast, cause and effect
Lab Materials: celery stalk   colored water   clear glass
Paper Handouts: Lab Book   Lab Record Cards   Lab Log   a copy of Lab Graphic 9-1
Graphic Organizer: Glue Lab Graphic 9-1 on the right pocket of the Lab Book
for this lab.
Concept: Vascular plants move water and nutrients through their parts using tube-like structures called vascular bundles.
Research: Read *Lots of Science Library Book #9* and review vascular systems in plants.
Procedure: Cut the celery stalk at an angle and place it in the water. Observe the celery in two hours and note any changes. At the end of two days, take the celery out of the water and cut it about an inch from the bottom. Observe the vascular tubes that carry the colored water up the celery stalk. Mark your Lab Log for two days.
Observations: Check the lab daily. How does the water move through the celery stalk? How far up the celery stalk did the water move?
Record the Data: Each day on the Lab Record Cards record Lab 9-1, the date, and a sketch of the celery stalk.
Conclusion: What conclusion can you draw about vascular plants from this lab?
Communicate the Conclusion: On a Lab Record Card, compare vascular plants to non-vascular plants (bryophytes). Include the effect of the vascular system on the characteristics of these types of plants.
Questions: List questions sparked by this lab.
New Loop: Choose a question to investigate further.

Find a Fern - Plant ID Book

Materials: Nature Guide Book
Paper Handouts: 8.5"x11" sheet of white paper   Plant ID Book
Graphic Organizer: Make a Large Question and Answer Book. Glue it side-by-side to the student's Plant ID Book.
Students select two fern plants to feature in their Plant ID Books. Draw one plant on each tab. Record the plant observations under the tab.

Continue with the Reproduction Pocket Book made in Lesson 5 by asking your students to make a card for ferns. Put it in the sexual reproduction pocket of the book.

Experiences, Investigations, and Research

Select one or more of the following activities for individual or group enrichment projects. Allow your students to determine the format in which they would like to report, share, or graphically present what they have discovered. This should be a creative investigation that utilizes your students' strengths.

 1. Fill a tiny clay pot with peat moss and turn it upside down in a dish of water. Sprinkle fern spores collected from the underside of fern leaves on the outside of the clay pot. Place a glass jar over the whole thing. Watch the fern plants develop over a period of several weeks.

2. Investigate why Azolla ferns are grown in rice fields in Southeast Asia.

3. Discover what part ferns played in the formation of coal deposits used today. Investigate how and why this happened and its impact on our world.

http://amer
fernsoc.org

Notes

What are gymnosperms?

Plant Concepts:

- Gymnosperms are vascular plants.
- Gymnos is a Greek word meaning "naked" and sperma means "seed."
- A gymnosperm seed is not in a case.
- Gymnosperms produce seeds but not true flowers.
- They are woody plants in the form of trees, shrubs, or vines.
- Most gymnosperms are evergreen and have needles or scalelike leaves and deep roots.
- Four orders are cycads, ginkgophytes, gnetophytes, and conifers.

Vocabulary Words: gymnosperm woody plant tree shrub vine
evergreen needle scalelike *cycad (SY kad) *ginkgophyte (GING ko fite)
*gnetophyte (NET oh fite) *conifer (KON uh fer)

Read: *Lots of Science Library Book #10.*

Activities:

Gymnosperms – Graphic Organizer

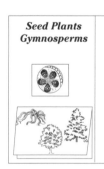

Seed Plants Gymnosperms

Focus Skill: explaining the characteristics of a group
Paper Handouts: 8.5"x11" sheet of paper   a copy of Graphics 10A – B
              Gymnosperm Book made in Lesson 6
Graphic Organizer: Make a Half Book. Glue Graphic 10B on the cover
       and title the book.

✎ Go on a nature walk and collect gymnosperm needles and leaves.
   Inside the book, use 2" tape to display them.

✎✎ Inside the book, write clue words about gymnosperms: *vascular plants, naked
   seeds, no true flowers, woody plant.* Write one or two complete sentences describing
   gymnosperms. Compare and contrast two different species of gymnosperms growing in
   your area.

✎✎✎ Complete ✎✎. List the four orders of gymnosperms. Give examples for each. Include
   illustrations if possible.

   Glue this Half Book onto the bottom of the left page in the Gymnosperm Book.
   Glue the 'vascular Graphic' (10A) to the top of that page. Label the graphic *vascular plant*.
   Store for future use.

Leaf Collection

Begin a leaf collection to be used in Lesson 17. The leaves may be kept in a folder or plastic bag.

Gymnosperm Hunt – Plant ID Book

Materials: Nature Guide Book
Paper Handouts: 8.5"x11" sheet of white paper   Plant ID Book
Graphic Organizer: Make a Large Question and Answer Book. Glue it side-by-side to the Plant ID Book.
Find plants with the characteristics of gymnosperms.
Remember, most gymnosperms bear cones. Draw the plant and a close-up of the leaf. Write about the plant under the tab.

Experience, Investigations, and Research

Select one or more of the following activities for individual or group enrichment projects. Allow your students to determine the format in which they would like to report, share, or graphically present what they have discovered. This should be a creative investigation that utilizes your students' strengths.

Geography

1. The ginkophyte plant is no longer found in the wild. These plants are grown by gardeners primarily in China and Japan. Research the ginkophyte plant and discover how it is used.

2. Research ancient gymnosperms and what they tell us about the past.

What is a conifer?

Plant Concepts:

- Conifers are the largest and most common order of gymnosperms.
- Conifers bear cones.
- Male cones produce pollen. They are smaller than female cones.
- Female cones produce seeds. They are formed in scales of the cone.
- Fertilization occurs when wind carries pollen to the seeds.
- Most conifers carry out photosynthesis all year round.
- Conifers are called evergreens because most of them stay green all year long.
- They lose their leaves, but not all at once.
- Conifer leaves are usually needle, or scalelike, and vary in shape and size.

Teacher's Note: An alternative assessment suggestion for this lesson is found on pages 78-79. If Graphic Pages are being consumed, photocopy assessment graphics needed first.

Vocabulary Words: conifer   scales   cones   male   female   clusters   ripe   evergreen   *ovule (OH vyool)   *pollen   *softwood

Read: *Lots of Science Library Book #11.*

Activities:

Conifers are Gymnosperms – Graphic Organizer

Focus Skill: explaining a process
Paper Handouts: 8.5"x11" sheet of paper   a copy of Graphics 11A – C
               Gymnosperm Book
        ✎ brown and white construction paper

Graphic Organizer: Make a Half Book. Turn the book so the fold is on the left side.
    Glue Graphic 11A to the cover and title it. Inside, glue Graphic 11B.

✎ From brown construction paper, cut out 20 circle or heart shapes. From white construction paper, cut out 10 small oval shapes. Glue the brown shapes in an overlapping manner to create a pine cone look. Under some of the brown shapes, insert a white oval to represent the seeds that the cone holds.

✎✎ Write clue words about the reproduction of conifers: *male cones – small and produce pollen; female cones – larger and produce seeds.* Explain how the pollen from the male cones reaches the female cones.

✎✎✎ Complete ✎✎. Explain what happens after germination. Use your vocabulary words.

Glue the Half Book onto the right page of the Gymnosperm Book.
Glue the 'seeds graphic' (11C) to the top of that page.

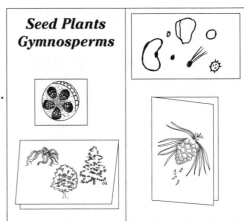

Seed Plants Gymnosperms

A Closer Look at a Cone

Activity Materials: pine cones   magnifying glass
Activity: Find several pine cones. Observe the outside of the cones. Some may have scales tightly packed and others may be more open. The cones with open scales have ripened seeds. The cones with tightly packed scales have seeds that are not ripe.

Open some of the pine cones and look for their seeds. These seeds are called pignolia. (peen yo lee uh)

Classifying Conifer Leaves and Cones – Plant ID Book

Materials: Nature Guide Book
Paper Handouts: 8.5"x11" sheet of white paper   Plant ID Book
Graphic Organizer: Make a Large Question and Answer Book. Glue it side-by-side to the Plant ID Book.
Students select two conifers to feature in their Plant ID Books. Draw one plant on each tab. Record the plant observations under the tabs.

Experiences, Investigations, and Research

Select one or more of the following activities for individual or group enrichment projects. Allow your students to determine the format in which they would like to report, share, or graphically present what they have discovered. This should be a creative investigation that utilizes your students' strengths.

 1. Make a cone bird feeder. Find a pine cone and tie string on one end of the cone. Spread peanut butter on the cone. Roll it in bird seed. Hang it on a tree. See who visits the pine cone bird feeder.

 2. Make a Coniferous Forest by folding paper into a Hot Dog. Cut along the fold. Fold one half into a Hamburger. Bring the edges up to the middle and crease (see diagram).
Draw a picture of a coniferous tree on the front being sure the tree overlaps the folds. Cut through all four layers. Open it up and color the trees. Stand up your coniferous forest. You may want to make more than one and tape them together.

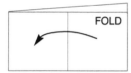 

 3. Begin researching coniferous forests. Use the world map found in the Graphics Pages. Find out their geographical locations, annual rainfall, and average temperature. Keep this information for a project that begins in Lesson 21.

 4. Look for pignolia seeds in your local supermarket. Make qualitative observations: shape, texture, color, and size.

 5. Find 10 objects in your home that originated from gymnosperms. Select one and investigate its production.

Plants Concept Map
Lessons 12-20

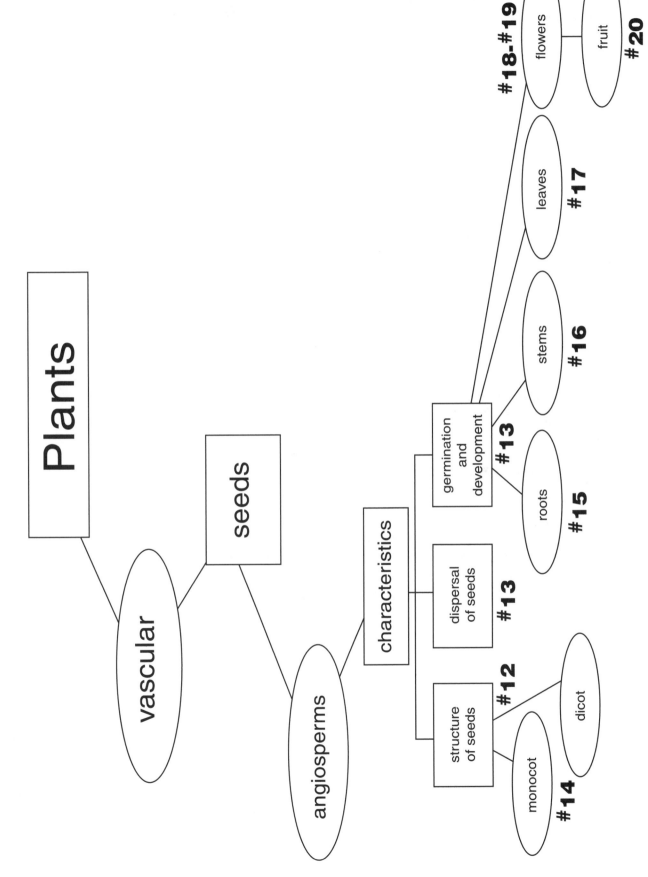

Plants

vascular

seeds

angiosperms

characteristics

structure
of seeds #12

dispersal
of seeds #13

germination
and
development #13

monocot #14

dicot

roots #15

stems #16

leaves #17

flowers #18-#19

fruit #20

What is an angiosperm?

Plant Concepts:

- Angiosperms have root systems, stems, leaves, and flowers.
- Angio is the Greek word for "covered," sperm means "seed."
- Angiosperms grow from seeds.
    - The parts of a seed are: seed coat, cotyledon, and embryo.
    - The seed coat protects the seed.
    - The cotyledon is the storage tissue that nourishes the embryo.
    - The embryo is the tiny plant inside the seed.
    - Over 70% of the food we eat comes from seeds.
    - Seeds are also used as spices and medicine.

Teacher's Note: It is not always easy to distinguish the fruit from the seed, as in the coconut and the cereal grains.

Vocabulary Words: seed   seed coat   embryo (EM bree o)
*cotyledon (kohte LEE don)   *testa (TES tah)

Read: *Lots of Science Library Book #12.*

Activities:

What is an Angiosperm? – Graphic Organizer

Paper Handouts: 12"x18" sheet of construction paper   a copy of Graphic 12A
Angiosperm Book made in Lesson 6

Graphic Organizer: Fold the paper in to a Hamburger and turn it so the fold is on the right. Glue the front of that book to the right of the Angiosperm Book. This will give you three pages in the Angiosperm Book.

Glue the 'vascular graphic' (12A) to the top of the left page.

Label the graphic *vascular plant.*

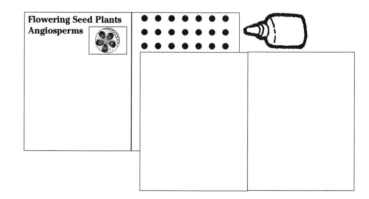

Investigative Loop – Inside a Seed – Lab 12-1

seed coat

embryo

cotyledon

Focus Skills: diagramming and labeling parts   qualitative observations
Lab Materials: paper towels   magnifying glass   four different types of seeds
Paper Handouts: Lab Book   a copy of Lab Graphic 12-1   Lab Record Cards
Graphic Organizer: Make another Pocket Book and glue it side-by-side to the
          Lab Book. Glue Lab Graphic 12-1 on left pocket.
Question: What is found inside a seed?
Research: Read the *Lots of Science Library Book #12* and review your knowledge of seeds.
Prediction: Predict what will be found inside a seed on a Lab Record Card. *"Inside the seeds are…"*
Procedure: Soak the four seeds in water overnight. Carefully open each one.
Observations: Observe the seed qualitatively before opening it. Look for the different parts inside
          each seed.
Record the Data: Label a Lab Record Card 12-1, the date, and diagram the inside of each seed.
          Label and describe each part.
Conclusions: Compare your observations with your predictions.
Communicate the Conclusions: On a Lab Record Card, write your conclusion about the insides of
          seeds.
Spark Questions: List questions sparked by this lab.
New Loop: Conduct an Investigative Loop to answer one or more of these questions.

Qualitative Observations of Seeds

Activity Materials: a variety of seeds (popcorn, lima beans, poppyseed, celery seed, or a 15
          bean soup mix)
Activity: Use qualitative observations to discover the qualities of the seeds: big, little, bumpy,
          round, pointed, color, smooth, rough, sticky, smell, etc. Ask each student to sort the seeds
          by their similarities. Seeds may be sorted in more than one way. Allow your student
          freedom to sort as he or she desires.

Experiences, Investigations, and Research

Select one or more of the following activities for individual or group enrichment projects. Allow
your students to determine the format in which they would like to report, share, or graphically
present what they have discovered. This should be a creative investigation that utilizes your
students' strengths.

1. Read *The Carrot Seed* by Ruth Krauss. ✎

2. Find approximately 20 seed pods of one kind (bean pods, pea pods, milkweed). Note
    the characteristics of the pods. Open them and count the peas inside. What did you
    find? How often did two peas in a pod occur? How often did three peas in a pod occur?

3. There is a banyan tree in India that has the largest spread in the world. This tree has
    more than 1,000 pillar roots, and more than 20,000 people can stand under it.
    Make a Record Holding Plants Journal out of a Bound Book. Add to it as you find new
    record breakers. Be sure to look through your *Lots of Science Library Books* to find
    data for your book.

How does an angiosperm develop and grow?

Plant Concepts:

- Seeds are scattered from parent plants to promote a better chance for survival.
- Seeds travel by wind, animals, water, and explosion.
- Seeds are dormant to complete their maturation process, to allow chemical changes to occur, and to allow their tough skins to soften.
- Germination is the process of an embryo sprouting and beginning to grow
- For germination to occur, water, oxygen, and suitable temperatures are required.

Vocabulary Words: explosion   mature   *dormancy   *germination

Read: *Lots of Science Library Book #13.*

Activities:

How Seeds Travel, Dormancy, and Germination - Graphic Organizer

Focus Skill: acquiring information
Paper Handouts: 5.5"x8.5" sheet of paper   a copy of Graphic 13A   Angiosperm Book
Graphic Organizer: Make a Hot Dog Book. Glue Graphic 13A to the cover. Cut
the top paper on the dotted lines.

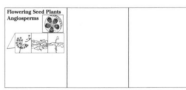

- ✎ Write/copy *seeds travel* on the first tab. Under the tab write/dictate
  clue words: *wind, animals, water, explosion.*
  Write/copy *seeds sleep* on the middle tab. Under the tab,
  write/dictate clue words: *need right conditions.*
  Write/copy *seeds grow* on the last tab. Under the tab,
  write/dictate clue words: *absorb water, begin to grow.*
- ✎✎ Label each tab: *Seeds Travel, Dormancy, Germination.* Write
  one sentence under each tab, describing the illustration on the front.
- ✎✎✎ Complete ✎✎. Select one plant and follow its development through these stages.
  Include the information about each stage under the tabs.
  Glue this book onto the middle of the left page in the Angiosperm Book.

Seed Travelers

Paper Handouts: a copy of Graphics 13B
Activity: Cut the piece of paper as indicated.
See illustrations for folding instructions.
Stand on a chair and drop your seed traveler. This is how a maple seed travels. Can you think of anything else that travels in this manner?
Now, cut off one of the blades and drop the seed traveler again. This is how a sycamore seed travels. Compare and contrast the way maple and sycamore seeds travel.

Focus Skill: cause and effect
Lab Materials: four lima beans   paper towels   clear glass jar
Paper Handouts: Lab Book   Lab Record Cards   Lab Log   a copy of Lab Graphic 13-1
Graphic Organizer: Glue Lab Graphic 13-1 on the right pocket Label of the Lab Book for this lab.
Question: How does a seed grow?
Research: Read the *Lots of Science Library Book #13* and review the germination
     process of seeds.
Predictions: On a Lab Record Card predict how the seed will grow. *"I think the seed will..."*
Procedure: Soak four lima beans in water overnight. Line the inside of a clear glass jar
     with damp paper towels. Place the seeds between the jar and the paper towels
     around the jar. Keep the paper towels damp.
Observations: Check the seeds daily and note changes. Mark your Lab Log for 10 days.
Record the Data: Each day label a Lab Record Card 13-1, the date, and describe the seeds.
     Include diagrams if desired.
Conclusions: Draw conclusions about the germination of a seed. Which part appears
     first? How do roots grow?
Communicate the Conclusions: In a Half Book, write a story or poem about the life of a seed,
     including the observations and conclusions from this lab. Be creative and illustrate your
     story. Fold the Half Book and store it in the Lab Book.
Spark Questions: Discuss questions sparked by this lab.
New Loop: When possible, design your own Investigative Loop for one of the questions.

Experiences, Investigations, and Research

Select one or more of the following activities for individual or group enrichment projects. Allow
your students to determine the format in which they would like to report, share, or graphically
present what they have discovered. This should be a creative investigation that utilizes your
students' strengths.

1. Find or purchase two plants in 2" pots. Leave one in the 2" container and replant the
   other into a large pot. Care for the plants in the same manner by placing them in a
   well-lit spot and water on a regular basis. Do not over water. Predict what will happen
   to each plant. After two weeks, or more, unearth the plants and compare the root
   systems. What relationship does the condition of the root system have with the
   growth of the plants?

2. Using the seed collection you began in Lesson 5, compare and contrast the
   qualities of the seeds in your collection. Sort the seeds according to:
   a) size – from the smallest to the largest seed
   b) color
   c) shape
   d) weight

3. Investigate the parent plants of the seeds in your collection. Research the
   seed producing organ of each plant, called a fruit. Sort the seeds in your
   collection by the method in which they are dispensed or scattered: wind,
   water, explosion, or animals.

Great Science Adventures

What are the two types of seeds?

Plant Concepts:

- There are two types of angiosperm seeds: monocot and dicot.
- Monocot comes from the Greek mono meaning "one" and cot is short for "cotyledon."
- Monocots have one cotyledon. Corn is a monocot seed plant.
- Dicot comes from the Greek di meaning "two" and cot is short for "cotyledon."
- Dicots have two cotyledons. A lima bean is a dicot seed plant.
- There are many more dicot plants than monocot plants.

Teacher's Note: An alternative assessment suggestion for this lesson is found on pages 78-79. If Graphic Pages are being consumed, photocopy assessment graphics needed first.

Vocabulary Words: monocot   dicot

Read: *Lots of Science Library Book #14.*

Activities:

Monocot Seed and Dicot Seed - Graphic Organizer

Focus Skill: compare and contrast
Paper Handouts: 8.5"x11" sheet of paper   a copy of Graphics 14A – D
            Angiosperm Book from the previous lesson.
Graphic Organizer: Make a Trifold Book. Fold it into a Hamburger. Cut on the fold to create two Mini-Trifold Books.
Monocot Seed: On one Mini-Trifold Book cover, draw/glue the outside of the corn seed, Graphic14A. Under the picture write *monocot*.   Open the Mini-Trifold Book. Cut off the third section. Draw/glue the inside of the corn seed, Graphic 14B, on the right section. Label the parts. Describe their functions on the left section of the book.
Dicot Seed: On the other Mini-Trifold Book cover, draw/glue the outside of the lima bean seed, Graphic 14D. Under the picture write *dicot*. Open the Mini-Trifold Book and draw/glue the inside of the lima bean seed, Graphic 14C, to the two left sections. Label the parts. Describe their functions on the right section.

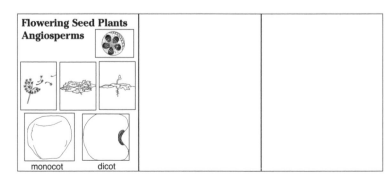

Materials: Nature Guide Book
Paper Handouts: 8.5"x11" sheet of white paper   Plant ID Book
Graphic Organizer: Make a Large Question and Answer Book. Glue it side-by-side to the Plant ID
    Book made in the previous lesson.
    Students select a monocot plant and a dicot plant to feature in their ID
    books. Draw one plant on each tab. Record the plant observations under the tabs.
    Monocot plants include lilies, tulips, grasses, chives, and pineapple.
    Dicot plants include marigold, hibiscus, and sunflower.

Experiences, Investigations, and Research

Select one or more of the following activities for individual or group enrichment projects. Allow your students to determine the format in which they would like to report, share, or graphically present what they have discovered. This should be a creative investigation that utilizes your students' strengths.

 1. Put on old, unwanted socks and walk through a field, vacant lot, or your backyard. Carefully remove them and use tweezers to collect any seeds attached. Observe the seeds. How did you identify them as seeds? Place them between wet paper towels to see if they will germinate.

 2. Carefully remove the socks, wet them. Place them in a pan or tray. Keep moist and watch the seeds germinate.

 3. Dig up dirt from a field and put it in a shallow pan. Water daily and see if any plants grow.

 4. Sort the seeds from your seed collection into two sets: monocots and dicots. Graph the results.

 5. Make a qualitative observation of your seed collection. Compare and contrast the external characteristics of the two sets of seeds. Make a 3 Tab Book to report your data.

 6. Make entries in your Record Holding Plants Journal.

http://www.urbanext.uiuc.edu/gpe/

What are roots and what do they do?

Plant Concepts:

- As seeds germinate, roots emerge first.
- Regardless of how a seed is placed in the ground, the roots grow downward because of gravity.
- Roots anchor plants, absorb water and minerals, and store food.
- Dicots, such as the carrot, usually have taproots.
- The taproot is a large main root with smaller roots branching out and down.
- Parts of the taproot include: root cap, epidermis, and root hairs.
- Monocots, such as grass, usually have fibrous root systems, which branch out from the stem.
- All roots contain xylem and phloem tissues. These vascular vessels carry nutrients throughout the plant.
- Xylem carries water and minerals up from the root to the rest of the plant.
- Phloem carries the food made in the leaves down to the roots.

Vocabulary Words: gravity   taproot   fibrous root   root cap   root hairs

*epidermis   *geotropism

Read: *Lots of Science Library Book #15.*

Activities:

Angiosperm Roots - Graphic Organizer

Focus Skill: compare and contrast
Paper Handouts: 2 sheets of 8.5"x11" paper
a copy of Graphics 15A – D
Angiosperm Book

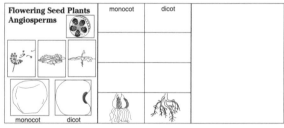

Graphic Organizer: Make two Small Question and Answer Books. Turn one with the fold on the left side and one with the fold on the right side. Glue the books next to each other on the middle page of the Angiosperm Book. On the top of the left book write *monocot*. On the top of the right book write *dicot*.

Monocot Roots: On the bottom left tab, glue/draw the fibrous root system, Graphic 15C. Open the tab. On the right section, glue/draw the identical picture, Graphic 15D. On the left section write about the parts and functions of the roots.

Dicot Roots: On the bottom right tab, glue/draw the taproot system, Graphic15B. Open the tab. On the left section, glue/draw the cross section of the taproot, Graphic 15A, and label the parts of it. On the right section, write about the parts and functions of the roots.

Functions of Roots – Graphic Organizer

Paper Handouts: 8.5"x11" sheet of paper   a copy of Graphics 15E – G
index cards

Graphic Organizer: Make a Pyramid Project. Glue/draw on each side pictures that illustrate the functions of roots. Describe the functions of the root on the index cards next to the Pyramid Project or hang them from each side as a mobile.

Collect Roots - Plant ID Book

Materials: Nature Guide Book

Paper Handouts: 8.5"x11" sheet of white paper   Plant ID Book

Graphic Organizer: Make a Large Question and Answer Book. Glue it side-by-side to the Plant ID Book made in the previous lesson.

Note: Be sure to have permission to dig up the plants before the activity begins. Students select two plants to feature in their Plant ID Books, examining the root structure of each. Draw one plant on each tab including the roots. Record the plant observations under the tabs.

Experiences, Investigations, and Research

Select one or more of the following activities for individual or group enrichment projects. Allow your students to determine the format in which they would like to report, share, or graphically present what they have discovered. This should be a creative investigation that utilizes your students' strengths.

 1.   Make Ginger Root Beer. Ingredients: 3 cups water, 2 inches finely grated fresh ginger, 2 tablespoon honey (or to taste), 1/2 sliced lemon. Mix the ingredients and boil for ten minutes. Strain the tea. Serve hot or iced.

 2.   Read *The Turnip* by Alexei Tolstoy. ✎   ✎✎

 3.   Investigate the orchid plant's aerial root system. This plant has its roots above the ground, absorbs water from the air, and minerals from decaying plant material. Discover other aerial plants.

 4.   Make " root" entries in your Record Holding Plants Journal.

What are stems and what do they do?

Plant Concepts:

- Stems usually grow above ground.
- Stems support leaves and flowers.
- They carry food and water up from the root to the leaves and rest of the plant (xylem).
- They carry food down from the leaves to the rest of the plant (phloem).
- Stems store food.
- Parts of a stem: epidermis, xylem, phloem.
- The epidermis is a protective covering of the stem.
- Vascular bundles consist of the xylem and phloem.
- Dicot plants contain a large number of vascular bundles, situated in a ring.
- Monocot plants contain a limited number of vascular bundles, situated in random order.
- Stems of dicot plants get thicker; monocot stems do not.
- Some underground stems are sweet potato and yam (tubers).

Vocabulary Words: vascular bundles   random   *epidermis

*herbaceous (hur BA shuhs)   *plumule (PLOOM yool)

Read: *Lots of Science Library Book #16.*

Activities:

Angiosperms - Graphic Organizer

Focus Skill: communicating information
Paper Handouts: a copy of Graphics 16 A – D
                 Angiosperm Book

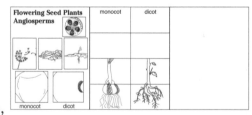

Graphic Organizer: Draw/glue the monocot stem picture, Graphic 16A, on the second tab from the bottom in the left book. Draw/glue the dicot stem picture, Graphic 16B, on the second tab from the bottom in the right book.
Under each tab, draw/glue the corresponding cross section of the monocot stem, Graphic 16C, and dicot stem, Graphic 16D.

✎ Draw arrows pointing up on one side of the stem and arrows pointing down on the other side of the stem indicating that water and nutrients move up and down the stems.

✎✎ Under each tab, explain the function of the stem and the arrangement of the vascular bundles.

✎✎✎ Under each tab, explain the function of the stem and describe how the stem of each type of plant grows, connecting that data to the arrangement of the vascular bundles.

Focus Skill: quantitative observations

Lab Materials: a white carnation   two glasses of water   knife or razor
       red and blue food coloring

Paper Handouts: Lab Book   Lab Record Cards   Lab Log   copy of Lab Graphic 16-1

Graphic Organizer: Make another Pocket Book and glue it side-by-side to the Lab
       Pocket Book. Glue lab Graphic 16-1 to the left pocket of the Lab Book for
       this lab.

Question: How far can water move up a carnation stem in five days?

Research: Read the *Lots of Science Library Book #16* and review what you know
       about stems.

Prediction: On a Lab Record Card, predict how many inches the water will move up the stem in
       five days.

Procedure: Slice the stem of the carnation a few inches up the stem. Fill both glasses 3/4
       full of water. Add four drops of blue food coloring to one glass and four drops of red food
       coloring to the other glass of water. Place 1/2 of the stem in the blue water and 1/2 of the
       stem in red water.

Observations: Daily observe the carnation. Note any changes. Mark your Lab Log for 5 days.

Record the Data: Each day, label a Lab Record Card with Lab 16-1, the date, and a description of
       changes observed in the carnation. Include the inches the colored water moved up the
       stem. Diagram the flower if desired.

Conclusions: Review your Lab Record Cards. Draw conclusions about how water is
       moved through a carnation. Note any different rates of movement during the lab.

Communicate the Conclusion: On a Lab Record Card, graph your data. Use the graph to explain
       this lab to someone who did not participate in it.

Spark Questions: Discuss questions sparked by this lab.

New Loop: Conduct research to answer one or more of these questions. When possible, design
       your own Investigative Loop.

Experiences, Investigations, and Research

Select one or more of the following activities for individual or group enrichment projects. Allow
your students to determine the format in which they would like to report, share, or graphically
present what they have discovered. This should be a creative investigation that utilizes your
students' strengths.

 1. Cut off a generous section of a potato including several eyes. Place it
     in a large glass jar with a lid. Cover the sides of the jar with black
     construction paper and screw on the lid. Check daily. After the leaves
     have sprouted, plant the potato in a pot or garden to watch it grow.

2. The stem of the papyrus plant was used by the ancient Egyptians and
     early Greeks as paper. Research how this was accomplished. Pretend to be an Egyptian
     reporter and describe the steps for papyrus paper making. Include illustrations.

 3. Cut the stem of a plant and use a magnifying glass examine the vascular bundle
     arrangement found there. Cut several other stems to examine their vascular bundle
     arrangements. Compare and contrast the various stems.

What are leaves and what do they do?

Plant Concepts:

- Three parts of a leaf are: blade, leaf stalk, and leaf base.
- The leaf blade is the main, flat part of the leaf.
- The leaf stalk is the part that connects the leaf to the stem.
- The leaf base attaches the leaf stalk to the stem.
- The epidermis protects the leaf.
- The mesophyll contains chlorophyll and helps in the exchange of gases.
- Food and water are carried by the vascular bundles, or veins.
- Monocot leaf veins are parallel.
- Dicot leaf veins are netlike.

Vocabulary Words: parallel   leaf blade   veins   leaf stalk   leaf base
*mesophyll (MEZ uh fill)   *lamina (LAM uh nuh)   *petiole (PET ee ole)   *cuticle (KYOO tih kul)

Read: *Lots of Science Library Book #17.*

Activities:

Angiosperm Leaves - Graphic Organizer

Focus Skill: describing common characteristics
Paper Handouts: a copy of Graphics 17A – D   Angiosperm Book
Graphic Organizer: Draw/glue the monocot leaf picture, Graphic 17A, on the second tab from the top in the left book. Draw/glue the dicot leaf picture, Graphic 17B, on the second tab from the top in the right book. Under the appropriate tab, draw/glue the monocot leaf, Graphic 17C, and dicot leaf, Graphic 17D.

✎ Observe the leaves of each type of plant and orally explain the similarities and differences in them. Outside, find a monocot leaf and a dicot leaf. Display each one under the correct tab with 2" tape.

✎✎ Under the tab, label the leaf parts and explain their functions. Write one descriptive phrase about each type of leaf.

✎✎✎ Complete ✎✎. Explain the function of each leaf part using the vocabulary words. Include the importance of leaves in photosynthesis learned in Lesson 2.

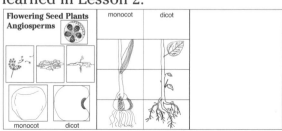

Investigative Loop – To Breathe or Not to Breathe – Lab 17-1

Focus Skill: drawing conclusions
Lab Materials: petroleum jelly   a growing plant
Paper Handouts: Lab Book   Lab Record Cards   Lab Log   a copy of Lab Graphic 17-1
Graphic Organizer: Glue Lab Graphic 17-1 on the right pocket of the Lab Book for this lab.
Question: What happens to a leaf that cannot absorb air?
Research: Read the *Lots of Science Library Books #1* and *#17* and review the needs of plants.
Prediction: On a Lab Record Card, record Lab 17-1 and write a step-by-step prediction of how the leaf will look in the next five days.
Procedure: Apply a generous amount of petroleum jelly on the top and bottom of one leaf of the growing plant.

Observations: Each day observe the leaf. Note any changes. Mark your Lab Log for 5 days.
Record the Data: Each day, label a Lab Record Card with Lab 17-1, the date and qualitative observations of the leaf. Diagram the leaf if desired.
Conclusions: Review your Lab Record Cards. Draw conclusions about the changes observed in the leaf. Why did the leaf change?
Communicate the Conclusions: On a Lab Record Card, write the conclusions about the leaf in this lab.
Spark Questions: Discuss questions sparked by this lab.
New Loop: Choose one sparked question to investigate further.

Leaf Hunt - Plant ID Book

Materials: Nature Guide Book
Paper Handouts: 8.5"x11" sheet of white paper   Plant ID Book
Graphic Organizer: Make a Large Question and Answer Book. Glue it side-by-side to the Plant ID Book.
Students select two plants to feature in their Plant ID Books, examining the leaf structure of each. Draw one plant on each tab, highlighting the leaves. Record the plant observations under the tabs.

Parallel and Netlike Veins

Paper Handout: 8.5"x11" sheet of paper
Activity Materials: leaf collection   ink pad
Graphic Organizer: Make a Large Question and Answer Book.
Find leaves with parallel veins and leaves with netlike veins in the leaf collection. Press the leaves on the ink pad. On one tab, place the leaf with the netlike veins on the paper and press.
Do the same on the other tab with the leaf with the parallel lines.
Look at the leaf prints and veins. Label the prints accordingly. Review the functions of the xylem and phloem.

Experiences, Investigations, and Research

Select one or more of the following activities for individual or group enrichment projects. Allow your students to determine the format in which they would like to report, share, or graphically present what they have discovered. This should be a creative investigation that utilizes your students' strengths.

 1. Observe the leaves forming on the potato started in Lesson 16.

 2. Make mint leaf candy. Check your local health food store for fresh mint leaves. Wash mint leaves and let them dry completely. Beat two egg whites until stiff. Flavor sugar with peppermint or spearmint oil. Coat the mint leaves with egg white, dip in flavored sugar and lay in a flat pan covered with wax paper. Bake in a 250° oven until dry. Enjoy!

 3. Research the sundew plant. Make a Half Book for the report. Use the picture on Graphics Page 17 or a format of your own choice.

 4. Investigate poisonous plants such as poison ivy and poison oak. Research how to identify them by their leaves. Describe the effect of their poison.

 5. Create a recipe for an edible leaf salad. Write the instructions in a traditional recipe format. Look at a cookbook to see how recipes are written.

 6. Make entries in your Record Holding Plants Journal.

Notes

What are flowers and what do they do?

Plant Concepts:

- The main purpose of a flower is to produce seeds for reproduction.
- A flower contains the male and female reproductive parts.
- Flowers consist of four organs arranged in whorls, or circles.
- Four whorls are: sepals, petals, stamen, and pistil.
- Sepals protect the bud and hold the flower in place.
- Petals attract animals.
- Stamens are the male reproductive part of the flower.
- Pistils are the female reproductive part.
- Dicot plants have flower parts in multiples of four or five.
- Monocot plants have flower parts in multiples of three.
- Composite flowers are made up of many smaller flowers.
- Most dicot flowering plants are composite flowers.

Teacher's Note: An alternative assessment suggestion for this lesson is found on pages 78-79. If Graphic Pages are being consumed, photocopy assessment graphics needed first.

Vocabulary Words: reproduction petals bud composite flower stamen pistil
*whorl   *sepal (SEE pul)   *disk florets   *ray florets   *gamete   *complete flower
*incomplete flower

Read: *Lots of Science Library Book #18.*

Activities:

Angiosperm Flowers – Graphic Organizer

Focus Skill: compare and contrast

Paper Handouts: a copy of Graphics 18B – E
Angiosperm Book
✎ colored construction paper

Graphic Organizer: Draw/glue the monocot flower picture, Graphic 18B, on the top tab in the left book. Draw/glue the dicot flower picture, Graphic 18C, on the top tab in the right book.

Under the monocot flower tab, draw/glue the cross section of the flower, Graphic 18D. Under the dicot flower tab, draw/glue the composite flower picture, Graphic 18E.

✎ Cut out petal shapes from the construction paper. Under the monocot flower tab, create a 3 or 6 petal flower on the left section. Under the dicot flower tab, create a 4 or 5 petal flower on the right section.

✎✎ Label the parts of the flower on the cross section. On the left section, write clue words about the function of the flower: *reproductive parts, produces seeds.* Write clue words about monocot flowers: *petals are in multiples of 3.* In the right book write clue words about composite flowers: *many small flowers that look like one.* Write clue words about dicot flowers: *petals are in multiples of 4 or 5.*

✎✎✎ Label the parts of the flower on the cross section. Compare and contrast the monocot and dicot flowers under the tabs. Describe composite flowers.

Composite Flower - Graphic Organizer

Paper Handouts: 8.5"x11" sheet of paper   colored tissue paper (optional), a copy of Graphic 18A

Graphic Organizer: Make a Trifold Book. Glue Graphic 18A on the cover.
Cut around the flower being sure to leave part of the folded sides uncut.
Tear off bits of tissue and twist them on the end of a pencil.
Dip it in glue and place it in the circle.
Continue until the circle is full of tiny flowers or disk florets.
Or, if you do not have tissue paper, draw small disk florets in the circle.
Open the Trifold and write on the middle section:

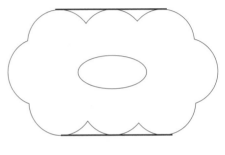

✎ *"A sunflower is made up of many flowers."*

✎✎ Through your observations and discoveries, justify this statement: "Almost all dicot flowering plants are composite flowers." See *Lots of Science Library Book #18*.

✎✎✎ In addition, list the names of composite flowers growing in your region. Use your vocabulary words.

Parts of a Flower - 3D Model

Activity Materials: modeling clay or dough   toothpicks
red and green construction paper

Activity: Make the stem out of clay and put a toothpick halfway down the middle of it as shown.
Cut the sepal out of green construction paper and put it on the stem.
Cut out the petals using red construction paper and put them on the stem.
Make the pistil out of clay and put it on top of the toothpick.
Make the stamens by putting bits of clay on the ends of toothpicks. Put the stamens in place as shown.

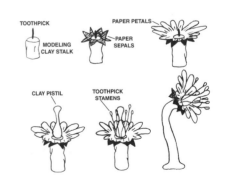

Materials: Nature Guide Book
Paper Handouts: 8.5"x11" sheet of white paper   Plant ID Book
Graphic Organizer: Make a Large Question and Answer Book. Glue it side-by-side to the Plant ID Book.

Students select two flowering plants to feature in their Plant ID Books. Draw one plant on each tab. Record the plant observations under the tabs.

Experiences, Investigations, and Research

Select one or more of the following activities for individual or group enrichment projects. Allow your students to determine the format in which they would like to report, share, or graphically present what they have discovered. This should be a creative investigation that utilizes your students' strengths.

 1. Carefully take a flower apart. Talk about the various parts. A tulip, petunia, snapdragon, daffodil, or morning glory works well for this activity.

 2. Mix equal parts of cornmeal and borax. Place in an airtight container. Put the flowers in the container. Cover the flowers, stem, and leaves gently with the mixture. Check every three days (petals feel like paper when dry). This generally requires from 5 days to 3 weeks, depending on the flower texture. Do not keep the flowers in the drying agent for too long. Petals will become brittle, and some flower color may be lost if the flowers dry too long. Once dry, gently remove flowers from mixture. Suggested flowers to use: carnation, cone flower, coralbells, daffodil, daisy, dogwood, gladiolus, hollyhock, lilac, magnolia, marigold, pansy, peony, rose, snapdragon, zinnia.

 3. Lay the flower flat between waxed paper. Put it between heavy books for 24 hours. Carefully remove the flower from the waxed paper and place it between paper towels or newspaper. Return it between the heavy books for several days. Use the flowers to make decorative pictures or greeting cards.

 4. The magnolia flower is a symbol for hospitality; find out why.

 5. Investigate common and uncommon edible flowers, such as cauliflower, artichoke, and petunias.

 6. Make candied flowers.
Ingredients and materials:
pesticide-free edible flowers (suggestions: pansies, Johnny-jump-ups, wild thyme, lavender, summer squash, calendula, carnation petals, dianthus, marigold, primrose, lilac, sage, nasturtiums, sweet violet, miniature rosebuds, rose petals)
one egg white
1 tsp. water
superfine sugar (the finer the crystals, the better)
paint brush
waxed paper

Beat egg white and water until almost meringue-like. Put a layer of sugar on a plate. "Paint" the flower or petals with the egg white mixture. Place the flower on the sugar. Sprinkle sugar on the flower and shake off the excess. Lay the flowers on waxed paper and place in an airy location to dry overnight. Or bake in the oven at a very low temperature keeping the oven door ajar. Store in an airtight container.

Notes

A GREAT LIFE SCIENCE STUDY

What is pollination?

Plant Concepts:

- Animals are attracted to a flower's nectar, colorful petals, and sweet smell.
- A flower contains male and female reproductive parts.
- Nectar is a sweet liquid inside flowers. Bees are attracted to it and use it to make honey.
- The stamen is the male reproductive part of the plant. It produces pollen.
- The pistil is the female reproductive part. It holds eggs.
- Pollination occurs when pollen from one flower falls on the pistil of another flower.
- Pollen grows a tube down the pistil to produce seeds.
- Pollen is moved from flower to flower by animals and wind.
- Wind-pollinated plants do not need a sweet smell or colorful petals.
- After a plant is pollinated, its seeds mature and are scattered by wind, animals, water, or explosion.

Vocabulary Words: nectar   stamen   pollen   pistil   ovary   style   pollination
*stigma (STIG muh)   *ovule   *immature

Read: *Lots of Science Library Book #19.*

Activities:

Pollination - Graphic Organizer

Focus Skill: sequencing
Paper Handouts: 8.5"x11" sheet of paper   Angiosperm Book
✎ a copy of Graphics 19B, D, E, F & G ✎✎ ✎✎✎ a copy of Graphics 19A – G
Graphic Organizer: Glue the 'seeds graphic' (19E) and the 'flower graphic' (19F) to the top of the last page in the Angiosperm Book.

✎ Make a Large Question and Answer Book. Draw/glue Graphic 19G on the cover. Inside, draw/glue Graphic 19B on the left tab and Graphic 19D on the right tab. Under the tabs, write/dictate clue words about pollination. Example: *honey bee gets nectar, pollen attaches to bee, bee goes to another flower, pollen rubs on the flower*

✎✎ Make a 4 Door Book. Draw/glue Graphic 19G on the cover. Draw/glue Graphics 19A-D in the correct order on the inside tabs. Write clue words for each step. Write one complete sentence for each step using the clue words.

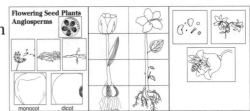

✎✎✎ Follow the directions in ✎✎ for making the book. Explain the pollination process, using your vocabulary words. Research other pollinators, such as bats and nocturnal moths, and explain the part they play in this process.

Glue this book in the middle of the last page in the Angiosperm Book.

Investigative Loop – Germinating Pollen Grains – Lab 19-1

Focus Skill: explaining a process
Lab Materials: various flowers with pollen   shallow dish   sugar
     sheet of glass   magnifying glass
Paper Handouts: Lab Book   Lab Record Cards   a copy of Lab Graphic 19-1
Graphic Organizer: Make another Pocket Book and glue it side-by-side to the Lab Book. Glue Graphic 19-1 on the left pocket in the Lab Book for this lab.
Concept: Pollen germinates in suitable conditions.
Research: Read the *Lots of Science Library Book #19* and review pollen.
Procedure: Dissolve a tablespoon of sugar in a cup of water. Pour the mixture into a shallow dish. Shake the pollen from the flowers into the solution. Cover with a sheet of glass. Keep it in a warm place for several hours.
Observations: Make qualitative observations at the beginning and end of this lab, using a magnifying glass.
Record the Data: On a Lab Record Card, record Lab 19-1, the date, and the changes observed in the dish during this lab.
Conclusions: Draw conclusions about the changes observed in this lab. How is the solution different? Why?
Communicate the Conclusions: On a Lab Record Card, explain the process of pollination, including observations and conclusions from this lab. Store the cards in the Lab Book.
Spark Questions: Discuss questions sparked by this lab.
New Loop: Conduct research to answer one or more of these questions. When possible design your own Investigative Loop.

Experiences, Investigations, and Research

Select one or more of the following activities for individual or group enrichment projects. Allow your students to determine the format in which they would like to report, share, or graphically present what they have discovered. This should be a creative investigation that utilizes your students' strengths.

1. Collect pollen from various flowers. Shake the pollen from each flower over separate pieces of black construction paper. Use a magnifying glass to observe the pollen. Compare and contrast the pollen. Cover the pollen with 2" clear tape to preserve it. Cut the paper to index card size and put it in the Lab Book.

2. Hand pollinate a flower. Pretend your finger is a bee. You land on a flower. Pollen from the male part (stamen) rubs off on you. Look at the pollen on your finger. You buzz around to another flower of the same kind and land on the sticky tip of the female part (pistil). You just pollinated a flower.

3. Observe how corn is planted in a field or in a garden. Investigate the cause and effect relationship of pollination to the development of an ear of corn.

4. Read *A Weed Is a Flower: The Life of George Washington Carver* by Aliki . ✎   ✎

5. Begin reading *The Forty Acre Swindle, George Washington Carver* by Dave and Neta Jackson.✎✎✎

What is a fruit?

Plant Concepts:

- The purpose of the fruit is to scatter seeds.
- After pollination, fruit develops around the seeds.
- Some seeds are covered with tough fiber; other seeds are covered with fleshy material.
- Some fruit trees shed their leaves; other fruit trees stay green all year.
- Structurally, some vegetables are actually fruits because they contain seeds.

Teachers' Note: An alternative assessment suggestion for this lesson is found on pages 78-79. If Graphic Pages are being consumed, photocopy assessment graphics needed first.

.

Vocabulary Words: fiber   fleshy   *ovules

Read: *Lots of Science Library Book #20.*

Activities:

Fruity Prints - Graphic Organizer

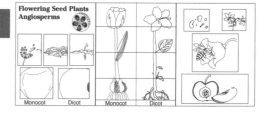

Focus Skill: explaining information

Activity Materials: ink pad or paper towels and tempera paint
various fruits such as pear, apple, etc.
vegetable fruits such as tomato, cucumber, squash

Paper Handouts: 8.5"x11" sheet of paper   newspaper

Graphic Organizer: Make a Half Book. Open the book and put it on a newspaper. Cut fruit in half. Make fruity prints on the cover and inside top of the book using the ink pad or tempera paint. Let it dry. Remove newspaper.

✎ Write/dictate the name of the fruit used in this activity. Circle the seeds in the fruity prints on the book.

✎✎ Write a simple paragraph on the inside bottom section of the book explaining how fruit develops and its purposes.

✎✎✎ Complete ✎✎. Research simple, aggregate, and multiple fruits and report the information orally or in the Half Book.

Glue this book on the bottom of the last page in the Angiosperm Book.

Binding the Types of Plants Book

Materials: Bryophyte Book   Fern Book   Gymnosperm Book   Angiosperm Book
a 12"x18" sheet of construction paper
All four plant books are completed. Discuss how to classify these groups: vascular types, seed plants, flowering plants. Glue the books side-by-side in the order in which they were completed. Make an outside cover with the 12"x18" paper, title, and decorate it.

Experiences, Investigations, and Research

Select one or more of the following activities for individual or group enrichment projects. Allow your students to determine the format in which they would like to report, share, or graphically present what they have discovered. This should be a creative investigation that utilizes your students' strengths.

 1. Make a fruit salad and invite your friends for the tasty treat. Suggestions: diced apples, bananas, oranges, cantalope, grapes, other seasonal fruits. Add sugar or honey to taste.

 2. Read *Stone Soup* by Marcia Brown. Make your own fruit and vegetable soup. Suggestions: squash, zucchini, tomatoes, okra in a broth and season to taste.

 3. Research the Supreme Court decisions in 1893 and 1980 that ruled the tomato must be considered a vegetable for the purpose of trade.

 4. A banana is 75% water. Research the water content of other fruits. Graph your results.

 5. Investigate the historic and geographic movement of fruits by explorers, settlers, and traders.

 6. Research scurvy. Explain what part it played in exploration.

 7. Read *The Story of Johnny Appleseed* by Aliki.

 8. Research Johnny Appleseed or read *Trail of Apple Blossoms* by Irene Hunt. Make a 4 Door Book and label the doors: *Who, What, When* and *Where*. Write the appropriate information about Johnny Appleseed under each door.

Notes

Plants Concept Map
Lessons 21-24
Numbers Refer to Lesson Numbers

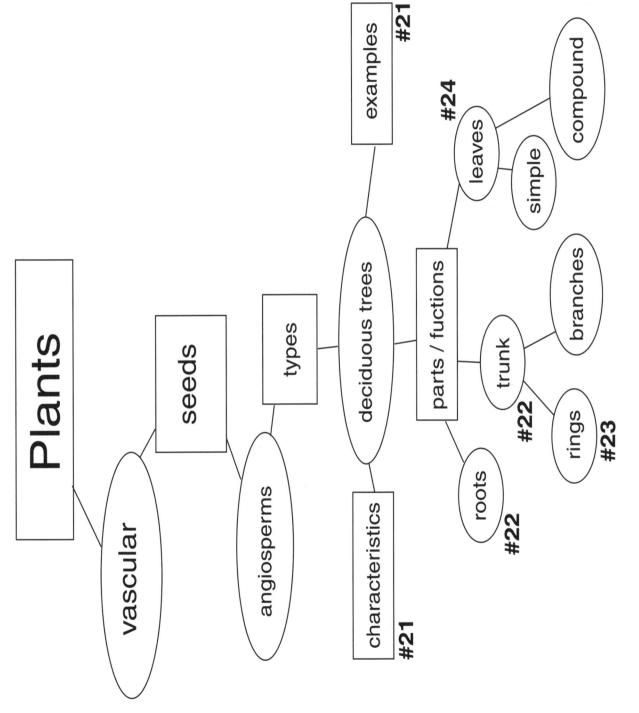

Plants

vascular

seeds

angiosperms

types

deciduous trees

examples
#21

characteristics
#21

parts / fuctions

leaves
#24

simple

compound

trunk
#22

branches

rings
#23

roots
#22

What are deciduous trees?

Plant Concepts:

- A tree has a woody stem (trunk), upper branches, and a crown.
- Most angiosperm trees are called deciduous, hardwoods, and broad leafed trees.
- Deciduous forests are located in various parts of the world.
- Some deciduous trees are maple, birch, oak, beech, and sycamore.
- Deciduous trees produce flowers and seeds in the form of a nut or fruit.

Vocabulary Words: trunk   hardwood   broad leaved   crown
*deciduous (dih SIH joo us)

Read: *Lots of Science Library Book #21.*

Activities:

Coniferous and Deciduous Trees - Graphic Organizer

Focus Skills: research and map skills
Paper Handouts: 12"x18" sheet of paper   a copy of Graphic Page 21
            a copy of the world map in the Graphics Page section
Graphics Organizer: Make a Shutter Fold Book. Cut the Graphics Page on Cut #1.
    Glue the bark picture on the cover of the Shutter Fold Book. Cut on the dotted lines. This
    will create a tab on each side of the front cover.

    On the top left side of the cover, write *Coniferous Trees*.
    On the top right side of the cover, write *Deciduous Trees*.

    Conduct research, document your findings, and report on the following:
    Average annual temperatures, high and low, average annual precipitation, and geographic
    locations of coniferous forests and deciduous forests.
    Record annual temperatures on the thermometers, high
    in one color and low in another color, record rainfall
    on the rain gauge graphics, and color the small
    world maps to indicate geographic locations. Glue
    these graphics inside the Shutter Fold Book, behind
    the appropriate shutter.

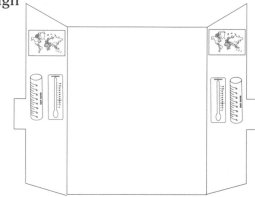

Tree ID - Plant ID Book

Materials: Nature Guide Book
Paper Handouts: 8.5"x11" sheet of white paper   Plant ID Book
Graphic Organizer: Make a Large Question and Answer Book. Glue it
side-by-side to the Plant ID Book.
Students select two deciduous trees to feature in their Plant ID
Books. Draw one tree on each tab. Record the plant observations.

The Completed Plant ID Book

Paper Handouts: Plant ID Book   11"x17" or 12"x18" sheet of paper
Graphic Organizer: Measure the Plant ID Book from the front right edge to the
back edge. Measure it from top to bottom. Cut a piece of paper to
the measurements. Glue the paper onto the Plant ID Book as a
cover. Title and decorate it.

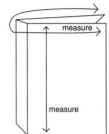

Experiences, Investigations, and Research

Select one or more of the following activities for individual or group enrichment projects. Allow your students to determine the format in which they would like to report,
share, or graphically present what they have discovered. This should be a creative investigation
that utilizes your students' strengths.

1. Find an area with plenty of trees. Blindfold a student and lead him/her to a tree. Hug
    the tree to feel how big it is. Place his/her cheek against the trunk to feel the texture of
    the bark. Smell the tree and feel its leaves. After the student is acquainted with the tree,
    lead him/her away from it. Turn the student around and remove the blindfold. Ask the
    student to find the tree.

2. Make a deciduous forest by folding a sheet of paper into a Hot Dog. Cut on the fold.
    Fold one half into a Hamburger.
    Fold both ends back to touch the mountain top. (see diagram)
    Draw a picture of a deciduous tree on the front. Be sure the tree overlaps the folded
    edges.
    Cut through all four layers, being sure not to cut off all the folded areas on either side.
    Open it up and color the trees. Stand your deciduous forest.
    You may want to make more than one and tape them together.
    Compare and contrast this forest with the coniferous forest made in Lesson 11.

3. Observe the trees in your area. Which trees do you think are deciduous trees and
    which ones are conifers? What characteristics that you learned about trees help you
    know how to identify each one? Use your Nature Guide Book to check your
    conclusions.

4. Do some trees provide cooler shade than others? Design an Investigative Loop to
    answer this question. Compare the shade of deciduous trees and the shade of
    coniferous trees.

5. Explain why shade trees are planted around homes and buildings.

What do tree trunks and roots do?

Plant Concepts:

- A tree has a woody stem called a trunk.
- The wood of a tree consists of a bark, cambium, and heartwood.
- Bark can be thick or thin. Bark protects the trunk.
- Trees grow and thicken in the cambium of the trunk.
- As new layers grow, old layers die. These dead layers support the tree in the heartwood of the trunk.
- Millions of short-lived root hairs absorb water.
- Young roots absorb nutrients and old roots anchor the tree.

Vocabulary Words: woody stem   trunk   extreme   fungi   bark   exchange   layer
core   heartwood   *cambium (KAM bee um)   *fissures (FISH ers)

Read: *Lots of Science Library Book #22.*

Activities:

Coniferous and Deciduous Trees - Graphic Organizer

Paper Handouts: 2 sheets of white paper   Trees Graphic Organizer from Lesson 21

Activity: Find a deciduous tree. Hold the white paper on the bark of the tree and rub a crayon over the paper. This is a bark rubbing of a deciduous tree. Do the same on a coniferous tree.

Graphic Organizer: Cut a piece of each bark rubbing and glue it under the graphics in the Trees Graphic Organizer.

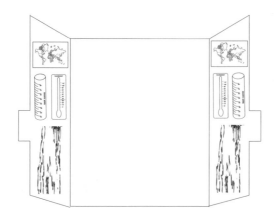

Focus Skill: explaining functions

Paper Handouts: 8.5"x11" sheet of paper   a copy of Graphic 22A   Trees Graphic Organizer

Graphic Organizer: Make a Half Book. Fold it into a Hamburger. Cut on the fold to create two Single Tab Books. Use one in this lesson and one in Lesson 23. Draw/glue Graphic 22A on the tab of one Single Tab Book.

✎ Color the bark of the tree dark brown. Color the rings light brown.

✎ ✎ Label the parts of the trunk on the front graphic. Inside the book, write clue words for each part of the trunk: *bark – protects trunk; cambium – growing part of the trunk; heartwood – dead wood, supports tree.*

✎ ✎ ✎ Complete ✎ ✎. Explain the function of each part of the trunk. Describe how the phloem and the xylem function in tree trunks.

Glue this book onto the top middle section inside the Trees Graphic Organizer.

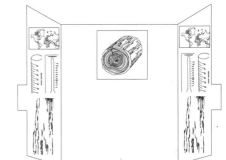

Bark Mold

Activity Materials: oatmeal box with lid   petroleum jelly
                        modeling clay   plaster of Paris

Activity: Put the lid on the oatmeal box and cut it in half lengthwise. Rub petroleum jelly on the inside of one half of the oatmeal container and put the box aside. Form the modeling clay into a shape about 4" x 4", about 1/2 inch thick. Apply petroleum jelly on one side of the modeling clay. Press the clay onto the bark of a tree, petroleum jelly side against the tree. Remove the clay and place it in the oatmeal container, petroleum jelly side up. Place the half lid on the end of the oatmeal container. This is a mold for the plaster of Paris. Mix plaster of Paris with water to get the consistency of pancake batter. Pour the plaster into the mold. When the plaster has hardened, remove the lid and carefully remove the plaster from the container. Paint the mold if desired.

Experiences, Investigations, and Research

Select one or more of the following activities for individual or group enrichment projects. Allow your students to determine the format in which they would like to report, share, or graphically present what they have discovered. This should be a creative investigation that utilizes your students' strengths.

1. Find several different species of trees and make bark rubbings. Make qualitative observations of each bark rubbing. Compare and contrast the findings. Take turns matching the rubbings to the trees.

2. Scientists have noticed that a tree's trunk is fatter at night and thinner during the day. Hypothesize why this might be true.

3. Read *A Tree Is Nice* by Janice May Udry.✎ ✎✎

4. Read *The Big Tree* by Bruce Hiscock.✎ ✎✎

What are tree rings and what do they tell us?

Plant Concepts:

- The cambium produces yearly rings.
- Thin-walled cells grow quickly in the spring, forming a light colored ring.
- Thick-walled cells grow slowly after summer, forming a dark colored ring.
- One pair of light and dark colored rings represent one year of growth.
- Tree growth stops in the fall of the year and resumes the next spring.
- The tree grows a wider ring during a wet summer and a narrower ring during a dry summer.

Vocabulary Words: cross section   climate   diameter   *concentric   *earlywood *latewood

Read: *Lots of Science Library Book #23.*

Activities:

Tree Rings - Graphic Organizer

Focus Skill: inferring
Paper Handouts: Trees Graphic Organizer   a copy of Graphic 23A
                   Single Tab Book made in Lesson 22
Graphic Organizer: Draw/glue Graphic 23A on the tab of the Single Tab Book. Count
     the rings. How old is this tree? **Answer: 12 years old**
   ✎ Inside the book, begin with the center ring and draw concentric rings, trying
      not to touch another circle. Draw enough rings to make the tree as old as
      the student drawing it.
   ✎✎ Inside the book, write clue words about tree rings: *grows dark and light ring each year,
      rings tell life of tree, rings are in the cambium of the trunk.* Make up a story explaining
      the life of the tree on the cover by looking at the rings.
   ✎✎✎ Complete ✎✎. Research dendrochronology and Andrew Douglas.

   Glue this book onto the bottom middle section of the Trees Graphic Organizer.

Focus Skill: research
Paper Handouts: Trees Graphic Organizer
Graphic Organizer: On the middle section of the Trees Graphic Organizer, make a list of coniferous trees to the left of the Single Tab Books. Make a list of deciduous trees to the right of these books.

Use the front shutter tabs as a guide to cut two slits in the middle section, between the lists of trees and the Single Tab Books. Roll the Tree Shutter Folds towards the middle and insert the tabs into the cut slits. This gives the Graphics Organizer the 'look' of trees. Find leaves from each type of tree and tape them to the back so they stick out on the side of each trunk.

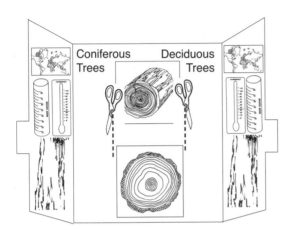 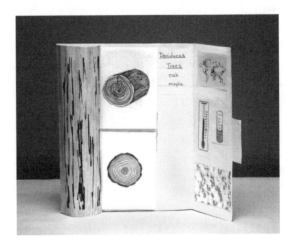

Experiences, Investigations, and Research

Select one or more of the following activities for individual or group enrichment projects. Allow your students to determine the format in which they would like to report, share, or graphically present what they have discovered. This should be a creative investigation that utilizes your students' strengths.

 1. Read *The Giving Tree* by Shel Silverstein. ✎   ✎✎

 2. Research trees that are used as state and country symbols. List them under the correct category of tree on your Coniferous and Deciduous Tree Graphics Organizer.

 3. Find or purchase firewood. Saw or sand off the end to clearly expose the tree rings. Examine the rings.
    How old was the tree when it was cut?
    Is there any other information the rings tell you?

 4. Begin reading *Miracles on Maple Hill* by Virginia Sorensen, a story about families working together at maple syrup collecting time. ✎✎✎

Why do deciduous leaves change color?

Plant Concepts:

- Simple leaves have one blade.
- Compound leaves have more than one blade.
- Most monocot plants have simple leaves.
- Dicot plants can have simple or compound leaves.
- A tree receives less sun and rain in autumn and winter, causing its food factory to shut down.
- A layer of cork forms at the end of each leaf stalk, stopping the water flow to the leaves.
- Leaves cannot make their own food without water and light.
- Chlorophyll breaks down and the other colors, such as orange and red, begin to show. These pigments are usually hidden by active chlorophyll.
- Eventually, the tree drops its leaves due to lack of water and food.

Teachers' Note: An alternative assessment for this lesson is found on pages 78-79. If Graphic Pages are being consumed, photocopy assessment graphics needed first.

Vocabulary Words: layer   cork   *simple leaf   *compound leaf

Read: *Lots of Science Library Book #24.*

Activities:

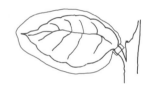

Investigative Loop – Trick a Leaf – Lab 24-1

Focus Skill: inferring
Lab Materials: a deciduous tree   opaque plastic bag
Paper Handouts: Lab Book   Lab Record Cards   Lab Log   a copy of Lab Graphic 24-1
Graphics Organizer: Glue Lab Graphic 24-1 on the right pocket in the Lab Book for this lab.
Question: What happens to a deciduous tree leaf that cannot receive air and sunlight?
Research: Read the *Lots of Science Library Book #24* and review deciduous tree leaves.
Predictions: Predict the changes that will be observable in the leaf when it cannot get air and
    sunlight.
Procedure: Cover a leaf on a deciduous tree with the plastic bag, tying it off with a rubber band at
    the stalk. Be sure you have permission to use the tree if it is not your own.
Observations: Observe the leaf daily. Note any changes. Mark the Lab Log for 10 days.
Record the Data: On a Lab Record Card, record Lab 24-1, the date, and observations of the leaf
    daily. Diagram the leaf if desired.
Conclusions: At the end of the lab, draw conclusions about the changes observed in the leaf. Why
    does the leaf look different?
Communicate the Conclusions: On a Lab Record Card, explain the conclusions from this lab.
    Describe the natural process that this lab represents.
Spark Questions: List questions sparked by this lab.
New Loop: Choose one sparked question to investigate further.

The Completed Lab Book

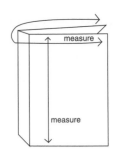

Paper Handout: Lab Book   11"x17" or 12"x18" sheet of paper
Graphic Organizer: Measure the Lab Book from the front right edge to the back edge. Measure it from top to bottom. Cut a piece of paper to the measurements. Glue the paper onto the Lab Book as a cover. Title and decorate it.

Leaves Change Color - Graphic Organizer

Focus Skill: sequencing a process
Paper Handouts: 8.5"x11" sheet of paper   a copy of Graphic 24A
Graphic Organizer: Make a Half Book. Draw/glue Graphic 24A on the cover.

- ✎ Color the cover leaf yellow. Draw falling leaves inside the book and color them brown, red, yellow, and orange. If appropriate, collect colored leaves and use 2" tape to attach them inside the book.
- ✎✎ Inside the book, write clue words explaining why leaves change colors and fall off deciduous trees: *fall and winter, less sun, less rain, cork forms on leaf stalk, factory shuts down, green color is gone, leaf falls.*
- ✎✎✎ Inside the book, explain why leaves change color and fall off deciduous trees. Explain how the loss of a tree's leaves affects the functions of that tree. Include photosynthesis, transpiration, respiration, and tropism.

Sort Leaves

Use the leaf collection or use Graphics 24B – E. Sort the leaves as simple or compound. Store the leaves in an envelope or make a Pocket Book for them.

Seasons of a Deciduous Tree - Graphic Organizer

Focus Skill: communicating information
Paper Handouts: 4 sheets of 8.5"x11" sheets of paper or a 12"x18" sheet of paper
                a copy of Graphics 24F – I   index cards
Graphic Organizer: Make 4 Pyramids from the 8.5"x11" paper. Glue them back to back to make a 4 Pyramid diorama. Color and glue each tree in a Pyramid, or make a 4 Door Book from the 12"x18" paper. Color and glue each tree on a tab inside the 4 Door Book. Decorate and title the cover.

- ✎ On index cards or under each tab, write/copy the season name for each tree. From catalogs, cut out pictures of clothes worn during that season and glue them on cards or under the tabs. Display the cards in front of each Pyramid for the diorama.
- ✎✎ Write the season of each tree under the tab or on index cards. Write a sentence on each one describing how trees in your region look in each season. Research and record the beginning and ending dates for each season in your region.
- ✎✎✎ Complete ✎✎. Describe what is happening to the tree internally during each season.

Experiences, Investigations, and Research

Select one or more of the following activities for individual or group enrichment projects. Allow your students to determine the format in which they would like to report, share, or graphically present what they have discovered. This should be a creative investigation that utilizes your students' strengths.

 1. Ask someone to describe a leaf from your collection using qualitative details: shape, size, texture, color, simple or compound, parallel or netlike veins, etc. See if you can recognize which leaf is being described. Take turns describing a leaf.

 2. Make a cardboard fence by cutting the top and bottom off a box. Place the fence under a deciduous tree. Monitor how many leaves fall in the space over the week. Predict how seasons will affect the results.

 3. Drop leaves of different shapes and note how they fall. How does the shape affect the fall? Do fresh green leaves and dead, dry leaves fall in the same manner?

 4. Read *Frog and Toad All Year* by Arnold Lobel.

Above and Beyond

Activities:

Plants We Eat Collage - Graphic Organizer

Paper Handouts: 8.5"x11" sheet of paper   a copy of Graphics 25 A – D
Graphic Organizer: Make a Small Question and Answer Book. Draw/glue the
     pictures of the fruit/flower, leaves, stem, and roots on each tab. Under the
     tab, write the foods we eat from each part of the plant.

Tons of Food - Graphing Information

Paper Handouts: a copy of Graphics 25 E
Activity: Give each student a copy of Graphic 25 E.
     Graph the amount of food eaten daily according to that information.
     Compare the amounts of food eaten daily. Discuss which ones are eaten by
     animals as well as people. Compare the world's consumption of types of
     food with your students' consumption of these types of food.

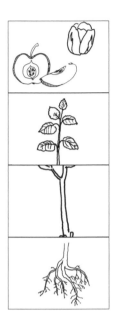

Experiences, Investigations, and Research

Select one or more of the following activities for individual or group enrichment projects. Allow your students to determine the format in which they would like to report, share, or graphically present what they have discovered. This should be a creative investigation that utilizes your students' strengths.

 1. Record the plants you eat every day for a week. Sort your list into groups: roots, stems, leaves, flower/fruit. Bar graph the information.

 2. Research one area of plant use: medicine, cosmetics, building materials, clothing and decorations. Report your findings.

 3. Visit a produce store. Look at different foods that come from plants and discuss them as parts of a plant.

 4. Read *Blueberries for Sal* by Robert McCloskey.✎ ✎✎

 5. Using all, or some, of the activities created in this program, make a cumulative project that can be used in a science presentation. Use *The Big Book of Books and Activities* by Dinah Zike for display ideas. Visit www.dinah.com or call 1-210-698-0123 for information on this book.

Assessment: An Ongoing Process

Students do not have to memorize every vocabulary word or fact presented in these science lessons. It is more important to teach them general science processes and cause and effect relationships. Factual content is needed for students to understand processes, but it should become familiar to them through repeated exposure, discussion, reading, research, presentations, and a small amount of memorization. You can determine the amount of content your students have retained by asking specific questions that begin with the following words: *name, list, define, label, identify, draw,* and *outline.*

Try to determine how much content your students have retained through discussions. Determine how many general ideas, concepts, and processes your students understand by asking them to describe or explain them. Ask leading questions that require answers based upon thought and analysis, not just facts. Use the following words and phrases as you discuss and evaluate: *why, how, describe, explain, determine,* and *predict.* Questions may sound like this:

What would happen if _____?   *Compare _____ to _____.*
Why do you think _____ happens?   *What does ___have in common with __?*
What do you think about _____?   *What is the importance of _____?*

Alternative Assessment Strategies

If you need to know specifically what your students have retained or need to assign your students a grade for the content learned in this program, we suggest using one of the following assessment strategies.

By the time your students have completed a lesson in this program, they will have written about, discussed, observed, and discovered the concepts of the lesson. However, it is still important for you to review the concepts that you are assessing prior to the assessment. By making your students aware of what you expect them to know, you provide a structure for their preparations for the assessments.

1)   At the end of each lesson, ask your students to restate the concepts taught in the lesson. For example, if they have made a 4 Door Book showing the steps of pollination, ask them to tell you about each step using the pictures as a prompt. This assessment can be done by you or by a student.

2) At the end of each lesson, ask your students to answer the questions on the inside back cover of the *Lots of Science Library Book* for that lesson. The answers to these questions may be done verbally or in writing. Ask older students to use their vocabulary words in context as they answer the questions. This is a far more effective method to determine their knowledge of the vocabulary words than a matching or multiple choice test on the words.

3) Provide your students with Plant Concept Maps partially completed. Ask them to fill in the blanks. Example for Lesson 3:

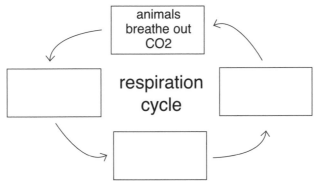

4) Use the 3D Graphic Organizers to assess your students' understanding of the concepts. Use the *Plant Concepts* listed on the teachers' pages to determine exactly what you want covered in the assessments. Primary and beginning students may use the pictures found on the Graphics Pages as guides for their assessments. By using the pictures, your students are sequencing and matching, while recalling information. Older students should draw their own pictures and use their vocabulary words in their descriptions of the concepts. Below are suggestions for this method of assessment.

   a) Lesson 2 – Use 4 pieces of paper to make a Layered Look Book. Your students will list the 8 steps of photosynthesis and draw/glue a picture for each step on these 8 pages. Older students will include vocabulary words in their descriptions. When grading this assessment, each description and drawing can be given up to 6 points, allowing 4 points for extra credit.

   b) Lesson 5 – Make a Large Question and Answer Book to describe the two ways that plants reproduce.

   c) Lesson 9 – Make a Shutter Fold Book. On and under one shutter write about bryophytes, their structure and general reproduction process. On and under the other shutter, write about ferns, their structure and general reproduction process. On the middle section, compare and contrast the two types of plant groups.

   d) Lesson 11 – Make a Large Question and Answer Book. On one tab draw/glue a gymnosperm and under it describe their common characteristics. On the other tab draw/glue the reproductive part of a conifer and describe how they reproduce.

   e) Lesson 14 – Make a Trifold Book. On the outside write about the seed and its parts. Inside, describe and diagram a monocot seed and a dicot seed. Compare and contrast these types of seeds.

   f) Lesson 18 – Make two Large Question and Answer Books and glue them side-by-side. Use each tab to describe the types and functions of roots, stems, leaves, and flowers.

   g) Lesson 20 – Make a Small Question and Answer Book. Use the four tabs to sequence the process of pollination.

   h) Lesson 24 – Make two Large Question and Answer Books and glue them side-by-side. Use each tab to diagram, label parts, and describe the function of the trunk, roots, rings, and leaves of deciduous trees.

Notes

Lots of Science Library Books

Each *Lots of Science Library Book* is made up of 16 inside pages, plus a front and back cover. All the covers to the *Lots of Science Library Books* are located at the front of this section. The covers are followed by the inside pages of the books.

How to Photocopy the *Lots of Science Library Books*

As part of their *Great Science Adventures,* your students will create *Lots of Science Library Books*. The *Lots of Science Library Books* are provided as consumable pages which may be cut out of the *Great Science Adventures* book at the line on the top of each page. If, however, you wish to make photocopies for your students, you can do so by following the instructions below.

To photocopy the inside pages of the *Lots of Science Library Books:*

1. Note that there is a "Star" above the line at the top of each *LSLB* sheet.

2. Locate the *LSLB* sheet that has a Star on it above page 16. Position this sheet on the glass of your photocopier so the side of the sheet which contains page 16 is facing down, and the Star above page 16 is in the left corner closest to you. Photocopy the page.

3. Turn the *LSLB* sheet over so that the side of the *LSLB* sheet containing page 6 is now face down. Position the sheet so the Star above page 6 is again in the left corner closest to you.

4. Insert the previously photocopied paper into the copier again, inserting it face down, with the Star at the end of the sheet that enters the copier last. Photocopy the page.

5. Repeat steps 1 through 4, above, for each *LSLB* sheet.

To photocopy the covers of the *Lots of Science Library Books:*

1. Insert "Cover Sheet A" in the photocopier with a Star positioned in the left corner closest to you, facing down. Photocopy the page.

2. Turn "Cover Sheet A" over so that the side you just photocopied is now facing you. Position the sheet so the Star is again in the left corner closest to you, facing down.

3. Insert the previously photocopied paper into the copier again, inserting it face down, with the Star entering the copier last. Photocopy the page.

4. Repeat steps 1 through 3, above, for "Cover Sheets" B, C, D, E, and F.

Note: The owner of the book has permission to photocopy the *Lots of Science Library Book* pages and covers for his/her classroom use only.

How to assemble the *Lots of Science Library Books*

Once you have made the photocopies or removed the consumable pages from this book, you are ready to assemble your *Lots of Science Library Books*. To do so, follow these instructions:

1. Cut each sheet, both covers and inside pages, on the solid lines.

2. Lay the inside pages on top of one another in this order: pages 2 and 15, pages 4 and 13, pages 6 and 11, pages 8 and 9.

3. Fold the stacked pages on the dotted line.

4. Turn the pages over so that pages 1 and 16 are on top.

5. Place the appropriate cover pages on the inside pages, with the front cover facing up.

6. Staple on the dotted line in two places.

You now have several completed *Lots of Science Library Books*.

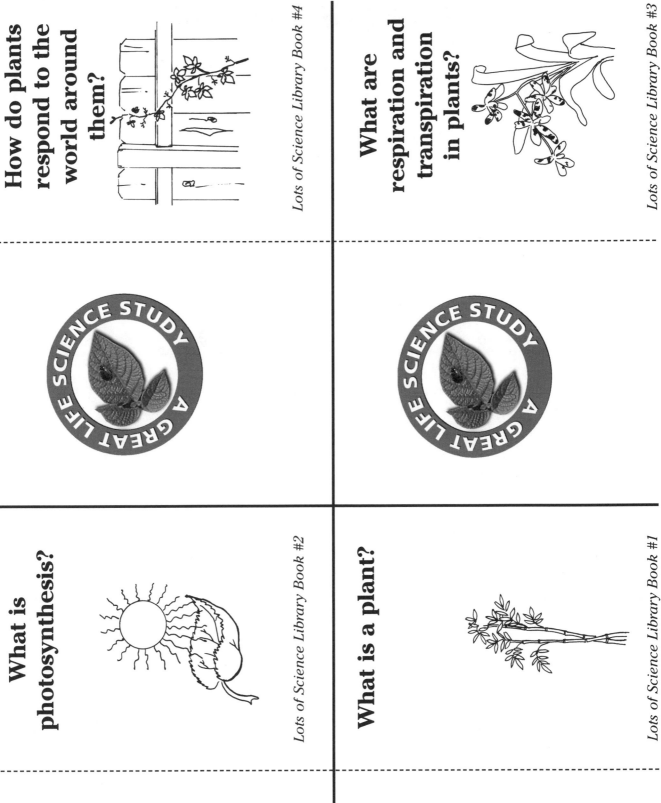

How do plants respond to the world around them?

Lots of Science Library Book #4

What are respiration and transpiration in plants?

Lots of Science Library Book #3

What is photosynthesis?

Lots of Science Library Book #2

What is a plant?

Lots of Science Library Book #1

SCIENCE STUDY
A GREAT LIFE

A

sensory
special
respond

* mobility
* stimuli
* tropism
* phototropism
* hydrotropism

inhale
exhale
oxygen

* carbon dioxide
* respiration
* transpiration
* stomata

How do plants
respond to water?
Why?

Why do you think
various plants
respond to touch?

How do plants and
animals help each
other in respiration?

Explain how plants
react to extra water.

green
absorbs
plant
sap
sugar
oxygen
hydrogen

* carbon dioxide
* chlorophyll
* photosynthesis

rooted
plant
air
cell
sunshine
grow
live

* reproduce
* biome

What is the
importance of
photosynthesis?

What does the word
photosynthesis
mean?

How are all plants
the same?

How do plants
differ?

How do bryophytes reproduce?

Lots of Science Library Book #8

What are bryophytes?

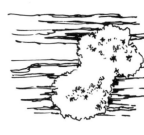

Lots of Science Library Book #7

What are the different types of plants?

Lots of Science Library Book #6

How do plants reproduce?

Lots of Science Library Book #5

B

spore
water
reproduce
spore case

* sporophyte

moss
simple
moist
cells
anchored

* bryophytes
* rhizoid

Explain how
bryophytes
reproduce.

Where are
bryophytes found?

Why do you think
they grow close to
the ground and live
in damp places?

moss
flower
classify
fern
seed

* botanist
* gymnosperm
* angiosperm

reproduce
cell

* gamete
* sexual reproduction
* asexual reproduction
* rhizomes
* zygote

Explain why
classifications
are important.

Explain how
botanists
classify plants.

Explain the two
ways plants
reproduce.

Describe the
ways plants
reproduce
asexually.

What is an angiosperm?

What is a conifer?

What are gymnosperms?

What are ferns and how do they reproduce?

seed
seed coat
embryo
* cotyledon
* testa

conifer
scales
cones
male
female
clusters
ripe
evergreen
* ovule
* pollen
* softwood

Describe angiosperms.

Compare angiosperms to gymnosperms.

Explain why conifers are called evergreens.

Where are the seeds of a conifer located?

woody
plant
tree
shrub
vine
needle
scalelike
evergreen
gymnosperm
* cycad
* ginkgophyte
* gnetophyte
* conifer

fern
water
tissue
stem
root
leaf / leaves
* xylem
* phloem
* vascular

Describe gymnosperms.

Explain why these plants are called gymnosperms.

Compare and contrast ferns to bryophytes.

What are stems and what do they do?

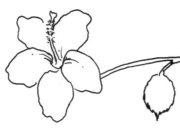

What are roots and what do they do?

What are the two types of seeds?

How does an angiosperm develop and grow?

SCIENCE STUDY
A GREAT LIFE

random
vascular bundles
* epidermis
* herbaceous
* plumule

gravity
root cap
taproot
root hairs
fibrous root
* epidermis
* geotropism

Explain the functions
of the stem.

Explain the functions
of roots.

monocot
dicot

mature
explosion
* germination
* dormancy

Explain how all
seeds are alike.

How do seeds differ?

Explain how and
why seeds are
scattered.

What is a fruit?

What is pollination?

What are flowers and what do they do?

What are leaves and what do they do?

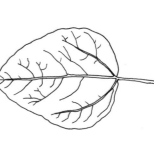

Describe the main purpose of the fruit.

fiber
fleshy

* ovules

What is the importance of fruit to its plant?

reproduction
petals
stamen
pistil
bud
composite
flower

* sepal
* disk florets
* ray florets
* complete flower
* incomplete flower
* gametes
* whorl

Explain the function of each part of a flower.

Describe the difference between a monocot flower and a dicot flower.

Explain how pollen is passed from flower to flower.

What is pollination?

nectar
stamen
pollen
pistil
pollination
ovary
style

* immature
* stigma
* ovules

Compare monocot leaves to dicot leaves.

Explain the function of the leaf's covering.

veins
parallel
leaf blade
leaf stalk
leaf base

* mesophyll
* lamina
* petiole
* cuticle

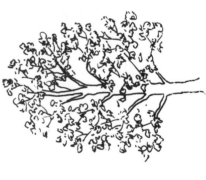

F

layer
cork

* simple leaf
* compound leaf

cross section
climate
diameter

* concentric
* earlywood
* latewood

What do deciduous
trees have in common
with conifers?

Explain why the leaves
of a deciduous tree
changes color.

Describe the information
tree rings tell us.

Explain how a tree ring
is formed.

---

woody stem
core
trunk
heartwood
extreme
fungi
bark
exchange
layer

* fissures
* cambium

trunk
hardwood
broad leaved
crown

* deciduous

Explain the function
of the roots on a
deciduous tree.

What is the importance
of the bark on a tree?

Describe how to
determine if a tree is
deciduous.

Explain why deciduous
trees are called
hardwoods.

Plants are found in all areas of the world. Plants can be found in the water, in a desert, and in the cold arctic.

Can you name plants in your region?

Can you name plants found in these biomes: desert, grasslands, rain forest, and tundra?

Plants become food for us to eat. Plants become our clothes.

Plants are used to make medicine, perfume, rubber, matches, houses, fabric, and much more.

All plants need air to live and grow.

Plants use carbon dioxide from the air to make their own food.

All plants need sunlight to live and grow.

Plants use sunlight and water in a unique way to produce food. See *Lots of Science Library Book 2.*

Did you see any plants today?

How did you know that what you saw was a plant?

Plants are all around us.

DID YOU KNOW?

Botany is the scientific study of plants. The word botany is Greek for "herb" or "plant."

FASCINATING FACTS

Some spruce trees found in the arctic areas of Canada take more than 100 years to grow just 20 inches.

All plants make new plants, or reproduce.

Not all plants reproduce in the same way as you will find out in *Lots of Science Library Book 5.*

Think of the plants you see every day. Can you name characteristics they all have in common?

Plants differ in many ways, but in some very important ways they are all the same.

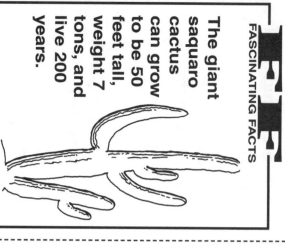

FASCINATING FACTS

The giant saquaro cactus can grow to be 50 feet tall, weight 7 tons, and live 200 years.

All plants need water to live and grow.

Most plants absorb water through their roots and lose it through their leaves.

Most plants are rooted in one location.

Plants can be annuals and live only one year. Plants of another species may live for thousands of years.

There are more than 300,000 different kinds, or species, of plants in the world.

About 90% of all plants produce flowers.

All plants grow throughout their lives.

Their rate of growth varies depending on how much light, water, and nutrients they receive. Plants also have special hormones that regulate their growth.

All plants contain more than one cell.

DID YOU KNOW?

Photo means "light," and synthesis means "putting together."

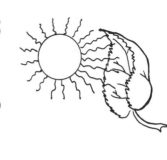

The chlorophyll captures, or absorbs, sunlight energy.

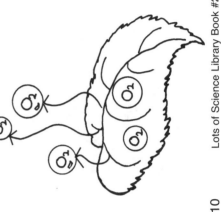

In most plants, chloroplasts are the tiny bodies in the plant cells that contain the chlorophyll.

The plant mixes the hydrogen and carbon dioxide to make sugar and other food. This food is called sap.

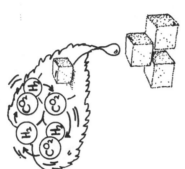

12

The oxygen is released into the air by the plant.

10

What do you do when you feel hungry? What would you do if you could not move around to get your food or call for someone to bring you food?
This is the situation a plant finds itself in when it needs food.

1

Do you remember the things that all plants need to live and grow?

water

air (CO₂)

sunlight

3

WHO'S WHO

Jan Ingenhousz was a botanist from the Netherlands who first discovered some of the stages of photosynthesis. He lived from 1730 to 1799.

16

Think about the factories that people use to manufacture things. They are usually big, noisy, and smell like smoke or chemicals.
Plants are busy making food to feed the world, and they do it without a sound.

14

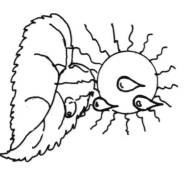

The plant mixes the
sunlight energy
with water.

Plants have
chlorophyll in the cells
of their leaves.

Most of the food is
manufactured in the
leaves of the plant.
Here's how the
factory
works:

The sunlight energy
splits the water into
hydrogen and oxygen.

The plant takes in
carbon dioxide
from the air.

This process of
food manufacturing
is called
photosynthesis.

Plants use these
three ingredients
to manufacture the
food they need to
live and grow.

All plants, from the tiny weed in
your driveway to the giant saguaro,
make their own food.

Plants have a
unique way
of getting
food. They
make it
themselves.

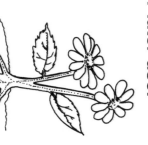

The plant moves
the sap throughout its
parts to use or
store for later use.

Plants are the first element
of every food chain on the
earth. Can you think of a
few examples of food chains?

All people and animals
live by either eating
plants or eating animals
that eat plants.

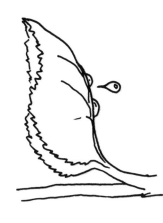

5

The plant releases oxygen in the process of making food. This adds oxygen to the air.

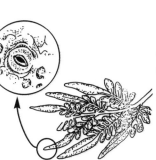

The respiration cycle of plants is the opposite of the respiration cycle of people and animals. This makes a perfect partnership for life on our planet.

7

When plants release extra water it is called transpiration.

Plants release millions of tons of water vapor into the air every day. Without transpiration animals could not live on this planet.

12   Lots of Science Library Book #3

Plants take in carbon dioxide through little holes, or pores, located on the underside of their leaves.

These pores are called stomata.

10   Lots of Science Library Book #3

Take a breath in and then let it out. Inhale again and then exhale.

How often do you do that in one hour? Count how many times you breathe normally in one minute and then multiply that by 60 minutes to find out how often you breathe in and out in one hour.

1   Lots of Science Library Book #3

If all the people and animals on Earth take in oxygen everyday, where does it all come from? In the same way, where does all the carbon dioxide go that is exhaled by all these people and animals?

3

Respiration is the opposite of photosynthesis. Compare these respiration facts with the facts you know about photosynthesis.

Respiration:

· uses oxygen to break apart sugar
· takes place in cells of living things
· takes place day or night
· releases energy

16   Lots of Science Library Book #3

If a plant releases more water than it takes in, it will wilt and then die.

This is the reason plants need more water in hot weather or when they are in direct sunlight.

14   Lots of Science Library Book #3

In a plant's respiration cycle, carbon dioxide gas is absorbed to make food. This takes carbon dioxide out of the air.

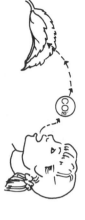

Did you know that an increased level of carbon dioxide in an environment can cause the temperature to rise?

6

DID YOU KNOW?

MORE FASCINATING FACTS

The tropical rain forest takes up about 6% of the land surface on Earth; but contains about 50% of Earth's plants. Scientists call the rain forests "the lungs of the earth."

Most plants have special systems that absorb water from the soil, move the water through their parts, and release any unused water into the air.

Extra water is released through the stomata on the underside of the leaves.

11

Without plants, life could not exist on this planet.

Without plants there would not be enough oxygen to inhale or food to eat.

9

All people and animals inhale oxygen and exhale carbon dioxide to live and grow. This process is called respiration.

Cells of all living things break down food so they can use its energy.

2

Review the steps of photosynthesis and see if you can answer the questions.

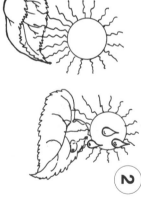

Review *Lots of Science Library Book #2* if you need help answering the questions.

4

FASCINATING FACTS

Desert plants do not transpire as much water as other plants. The giant saguaro cactus stores enough water to keep it alive three years.

15

FASCINATING FACTS

In one season, a corn plant will transpire 40 - 50 gallons of water.

13

Plants do not have the same special sensory cells that you have, so they do not react quickly to stimuli.

5

A plant can react to stimuli by moving its parts or by growing in a certain direction.

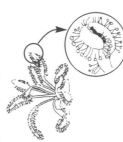

7

When a plant moves its parts or grows a certain way in response to stimuli, it is called tropism.

Some plants respond to touch. Have you ever seen a vine grow around a fence?

The response to touch in a plant is called haptotropism.

12   Lots of Science Library Book #4

Plants will grow in a certain direction to reach water.

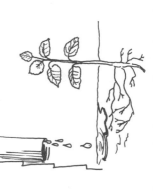

When a plant grows towards water, it is called hydrotropism. Remember, hydro means "water."

10   Lots of Science Library Book #4

Have you ever touched a hot stove? How did you respond? You probably pulled your hand away from the heat very quickly.

1

Lots of Science Library Book #4

Heat is a type of stimulus. Stimuli are things that you react to in a certain manner.

3

Stimuli is the plural word for stimulus. When you talk about one thing or type of thing you react to, it is called a stimulus.

---

16   Lots of Science Library Book #4

Think about their food, water, mobility, growth, air, and colors.

14   Lots of Science Library Book #4

Even though they do not move quickly, plants respond to certain stimuli around them.

Can you think of stimuli that your body reacts to at a slow rate? Examples: fighting disease, healing a cut, breaking down food.

Lots of Science Library Book #4

Plants grow toward sunlight.

Certain plants move or bend to receive more sunshine. When a plant responds to sunlight it is called phototropism. Remember, photo means "light."

Lots of Science Library Book #4

The mimosa pudica shrub is a sensitive plant. It closes its leaves when touched.

Ordinary bean plants fold their leaves very slightly at night.

Your body has special sensory cells that enable you to respond quickly.

These special cells, found in your nervous system, are called sensory neurons.

Lots of Science Library Book #4

If we compare plants and animals, plants might seem wimpy. Yet, plants are incredibly strong. Did you know that a dandelion can break concrete?

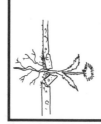

In the hot stove example, pulling your hand away from the heat quickly is called a response.

Humans have an elaborate nervous system that includes the brain, spinal cord, and millions of nerves.

Lots of Science Library Book #4

We know that plants respond to stimuli around them, but not like animals do. Can you think of any ways that plants and animals are alike? Can you think of ways that plants and animals differ?

Since this baby chick is made from the cells of two different animals, it is like both of them in various ways.

Some inherited characteristics are not easily seen, such as susceptibility to diseases.

5

Plants also reproduce. Their reproduction is similar to how dogs and chickens reproduce.

It is called sexual reproduction. Plants that reproduce in this manner use seeds and spores to reproduce.

7

Some plants grow new plants from their underground stems.

These stems are tubers, bulbs, or corms.

12   Lots of Science Library Book #5

Another method of reproduction used by some plants takes only one plant to make new plants.

It is called asexual reproduction or vegetative reproduction.

10   Lots of Science Library Book #5

All living things reproduce. Your biological mom and dad made another form of themselves when you were conceived.

Reproduction takes place at the smallest levels of life. Cells reproduce themselves. In fact, reproduction is an important characteristic of living things.

Lots of Science Library Book #5   1

Since the puppy is made from the cells of two different animals, he is like his dad and his mom in various ways.

Some characteristics may be easily seen, such as size, color, and markings.

3

FASCINATING FACTS

The underground stems of the papyrus plant were dried and used as firewood by the people in Ancient Egypt.

16   Lots of Science Library Book #5

Some plants send out new shoots from their roots.

The special roots sent out from the plant are called suckers. Banana trees reproduce new trees using suckers.

14   Lots of Science Library Book #5

WHO'S WHO

Gregor Mendel was an Austrian monk and botanist who discovered how characteristics are inherited by conducting thousands of genetic experiments using pea plants.

6

Lots of Science Library Book #5

In plants, a female sex cell and a male sex cell join together to create a new plant.

female sex cell + male sex cell = new plant

Plant sex cells are called gametes. When the male gamete and the female gamete fuse together it is called fertilization. Fertilization produces a zygote. The zygote is the first cell of the new plant.

8

Lots of Science Library Book #5

Since the new plant comes from a combination of the cells from two different plants, it is like both of the plants in various ways.

Plants that reproduce in this manner have the benefit of characteristics from two different plants.

9

Since there is only one parent plant, the new plants are exactly like the parent plant.

This can be a disadvantage if the parent plant has weak traits.

11

A puppy is made from the cells of a dad, or male dog, and a mom, or female dog. The puppy grows inside the female dog. He is born as a live animal.

2

Lots of Science Library Book #5

Some plants can grow new plants right on their leaves or from a broken-off leaf.

If you put an African violet plant leaf in a pot of soil with water, it will grow into a whole plant.

15

A baby chick is made from the cells of a male and a female, too. The female hen lays a fertilized egg. The chick grows inside the egg and when it is old enough it pecks its way out of the egg.

4

Lots of Science Library Book #5

Some plants have stems that run above or below the ground to make new plants.

Stems that grow along the ground are called runners. Stems that grow underground are called rhizomes.

13

There are about 300,000 kinds, or species, of plants in the world. Botanists classify these plants using two characteristics: their structure and how they reproduce.

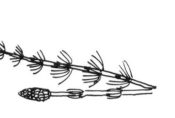

Using these questions as a guide, plants are divided into four main groups. The first group is the "moss-like" plants. There are about 10,000 kinds of moss-like plants.

These plants are bryophytes.

The third group is the nonflowering seed plant group. There are about 1,000 kinds of nonflowering seed plants.

This group is called gymnosperms.

12   Lots of Science Library Book #6

The second group is the fern plants. There are about 10,000 kinds of fern plants.

Lots of Science Library Book #6

Often things are sorted into groups. In the classified ads of a newspaper, ads are grouped so a person can easily find what they need. Vehicles are classified by how they look, the number of people they hold, and what they transport.

Lots of Science Library Book #6

Lots of Science Library Book #6

FASCINATING FACTS

Of all the types of plants in the world, about 90% are flowering seed plants. That means that there are about 270,000 species of angiosperms.

16   Lots of Science Library Book #6

The fourth group is the flowering seed plant group. There are about 240,000 kinds of flowering seed plants.

This group is called angiosperms.

14   Lots of Science Library Book #6

When botanists classify plants these questions are asked: How does the plant absorb water and food? How does the plant reproduce?

WHO'S WHO

Aristotle, a Greek philosopher and scientist, divided the living world into the categories of plants and animals. Aristotle lived from 384-322 B.C. and is considered the first great biologist.

The stem of a fern grows under the ground. The only part of the fern that can be seen above ground is the leaf.

stem

This group includes ferns, horsetails, and club mosses.

These "moss-like" plants are small and grow low to the ground. They usually live in very moist places.

The bryophyte group includes mosses, liverworts, and hornworts.

Look at these animals. How can you sort them? Where do they live? What do they eat? How do they look?

To sort, or classify, objects, common characteristics must be examined. What do these objects have in common?

The use of similar characteristics is used in all areas of science to classify living and nonliving things.

If you see a flower growing in your yard it is from this group of plants.

Angiosperms include all plants with flowers.

Most Christmas trees are gymnosperms.

You might think that all plants have roots, stems and leaves, but bryophytes don't. They have thread-like rhizoids instead of roots, short stalks instead of stems, and low spreading foliage.

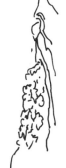

Since bryophytes move water and food slowly from cell to cell, they grow very low to the ground and live in damp places.

Bryophytes do not have an outer waterproof covering. Because of this they can easily dehydrate. This is another reason they must live in moist places.

Mosses provide shelter for small animals, and birds use them for nest building material. Mosses store minerals that other plants use when the mosses die, decay, and become part of the soil.

You can easily tell the difference between a moss and a liverwort. Mosses have an upright stalk with small, spirally arranged, leaf-like pieces attached to it.

Have you ever taken a walk through a shady, cool forest and found an area that seemed to be covered with a green velvety carpet? This carpet is actually a group of simple plants called mosses.

Mosses and liverworts are both in the group of plants called bryophytes, or "moss-like" plants.

Peat moss bogs are found in North America, Europe, and northern Asia. Research why they are located mainly in the Northern Hemisphere.

One type of moss is a swamp moss called peat. Peat moss is spongy and can absorb water. It is used by gardeners to enrich soil and hold water around plant roots.

Bryophytes are simple plants that absorb water and food from their environment through spaces between their cells. Think of a paper towel absorbing water, and you will have an idea of how these plants move water and food from cell to cell.

water

Bryophytes do not have true roots, stems, and leaves like many other plants.

Bryophytes have rhizoids that anchor them to the ground. Rhizoids are threadlike structures under the ground.

FASCINATING FACTS

Spanish moss, found hanging from trees in the southeastern United States, is not actually a type of moss. It is a flowering plant that is a member of the pineapple family.

Liverworts grow horizontally and have a flat, leafy appearance.

Liverworts are named this because they are shaped like a human liver. The word "wort" is an old English word for plant.

Mosses are found throughout the world, from the North Pole to the tropics. They are found on mountaintops or at the seashore. Mosses need a shady, moist area to grow. They are often found on rocks or trees.

Bryophytes are like all other plants. They need sunlight, water, and air. They also reproduce.

FASCINATING FACTS

Peat moss has antiseptic qualities and was once used as a dressing for wounds. Peat moss kills certain bacteria and fungi that cause diseases.

Mosses are often the first plants to appear after a volcanic eruption or forest fire. Since they grow low to the ground they also trap windblown soil. As soil accumulates, plants with roots can grow.

For this reason mosses are called "pioneer plants." As these pioneer plants die, they decay, beginning the formation of soil.

When the weather is very moist, the male sperm swims to the female egg and they join together.

male sperm

female egg

sperm + egg = zygote

zygote

The sperm is made to fit into the egg. When the gametes fuse, they form a zygote.

The sporophyte has a long stalk with a cup on the top of it.

FASCINATING FACTS

Some moss plants shoot spores into the air with such force that they sometimes land up to 6 feet away.

When the spores are mature, the case opens and the spores are released into the air.

Different plants have different ways of releasing their spores.

Bryophytes are small, simple plants that grow low to the ground in moist places. Mosses and liverworts are the most common bryophytes.

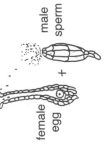

Bryophytes go through several steps before a new bryophyte can begin to grow into a mature plant.

FASCINATING FACTS

The spore cases of the stag horn clubmoss were once used in the production of fireworks because they produce a bright yellow powder.

Like all other plants, bryophytes reproduce. They have a unique way to produce new plants.

If you look closely at the top of most bryophytes you will see two different types of tops. One is male and produces sperm. One is female and produces eggs.

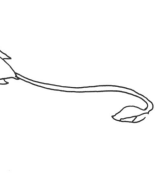

male

female

From this joining, a new plant, called a sporophyte, grows on the parent plant. The sporophyte does not look like the parent plant.

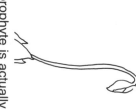

The sporophyte is actually a parasite, dependent on the parent plant for life.

The spore case is called a capsule.

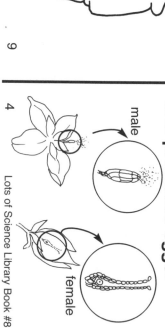

The cup looking object is a special case that holds spores made by the plant.

The spore cases of bryophytes are all different looking. Some of them have bright colors. Some of them are unusual shapes.

Wind can carry spores for long distances. When they land, spores can germinate and grow into new bryophytes.

These new plants produce male and female gametes so the process can continue.

Bryophytes are different from any other type of plant in the way they reproduce.

The process they use is called "alternation of generations."

There is an old saying, "a rolling stone gathers no moss." Based on what you have learned about bryophytes, what do you think that means?

Other parts of this system take the food leaves make through photosynthesis and move it through the plant.

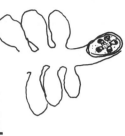

The phloem tissues carry the food from the leaves to the rest of the plant.

Remember that bryophytes move water and food in between their cell walls. They do not have a vascular system.

water

Bryophytes are called nonvascular plants.

These brown dots are spore cases. When the spores are mature, the cases open and spores are released.

When the spores ripen, the cells on the case become dry, curve back and break open.

12   Lots of Science Library Book #9

There are two types of leaves on a fern plant. One makes food for the plant through photosynthesis.

10   Lots of Science Library Book #9

Scientists study living ferns and fern fossils to learn more about these ancient plants.

1   Lots of Science Library Book #9

Ferns can grow tall and live in dry areas because they have a special system for moving water and food through their parts.

This system is made up of long tube-like cells that go throughout the plant.

3

From this joining, a new plant is produced. This new plant grows and produces spores so the cycle can begin again.

16   Lots of Science Library Book #9

The gametophytes produce male sex cells, or sperm, and female sex cells, or eggs.

female sex cells

male sex cells

14   Lots of Science Library Book #9

This manner of moving water and food through vascular tissues is used in all plants except bryophytes.

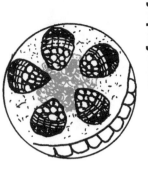

Plants that use the vascular tissues to move nutrients are called vascular plants. Ferns are vascular plants.

6

Lots of Science Library Book #9

---

stem

roots

The stem of a fern is called a rhizome.

8

Lots of Science Library Book #9

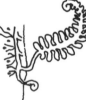

The second type of leaf makes food, and it also produces fern spores. These spores are found on the underside of the leaf and look like brown dots.

This type of leaf is called the fertile leaf. The brown dots are spore cases called sporangia.

11

When you look at a fern it is easy to think you are looking at a stem and leaves, but the stem of a fern is really underground.

The part of the fern that you see above ground is the leaf. The leaf is made up of tiny leaflets.

leaflet

9

The fern's leaves are called fronds. New leaves that develop on the plant are called fiddleheads. They look like the top of a violin when they begin growing.

Unlike bryophytes, ferns have true stems, root systems, and leaves. Roots anchor the plants and absorb water and nutrients from the soil.

2

Lots of Science Library Book #9

This special system of cells is called a vascular system. It takes in water through the roots and moves it throughout the whole plant.

The xylem tissues carry the water and nutrients from the roots to the rest of the plant.

4

When there is enough moisture, male sperm gametes swim to female gametes and join together.

As in most sexual reproduction processes in plants, the male gamete moves to reach the female gamete.

15

egg

sperm

Spores germinate and grow into very small plants called gametophytes.

The gametophytes are small, inconspicuous plants completely separate from the parent plant.

13

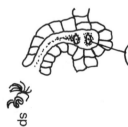

Most gymnosperms are called evergreens because they stay green all year.

Gymnosperms make up the oldest and largest trees such as the bristlecone pine and the giant redwoods.

Look at these examples of the four orders of gymnosperms.

2. gnetophyte

4. ginkophyte

1. cycad

3. conifer

Gnetophytes – This division of gymnosperm is a diverse group of plants. It consists mainly of shrubs, including the horizontally growing Welwitschia mirabilis.

12   Lots of Science Library Book #10

Ginkgophytes – This division of gymnosperm has only one species, the ginkgo or maidenhair tree. This order does not bear cones. Its leaves turn yellow and drop in the fall.

10   Lots of Science Library Book #10

Plants can be divided into two groups: vascular and nonvascular. Do you remember if ferns are vascular or nonvascular plants? Ferns are vascular plants.

1

Lots of Science Library Book #10

Although gymnosperms produce seeds, they do not produce true flowers.

3

There are fewer than 1,000 species of gymnosperms.

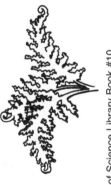

16   Lots of Science Library Book #10

14   Lots of Science Library Book #10

Cycads – These plants are tropical gymnosperms that look somewhat like ferns and palms. They are slow growing and usually produce one thick trunk and feather-like leaves.

FASCINATING FACTS

Since the seeds of the female ginkgo tree smell so bad (like spoiled butter) and can irritate people's skin, only male trees are planted along city streets!

FASCINATING FACTS

The custom of decorating Norway spruces and other conifers became widespread in the 19th Century.

Another type of vascular plants is gymnosperms. Gymnos is a Greek word that means "naked" sperma means "seed." A gymnosperm's seed is a naked seed, meaning it is not in a case.

Gymnosperms are woody plants. They are usually in the form of a tree, shrub, or vine.

Conifers – These gymnosperms include cypress, cedar, juniper, pine, fir, larch, cedar, spruce, hemlock, etc.

FASCINATING FACTS

The Welwitschia mirabilis is an unusual plant that grows in the deserts of southern Africa. It produces only two leaves that continue to grow throughout its life. It can live for more than 100 years. The leaves can grow to over 30 feet long. Wind and blowing sand shred the two leaves, giving an appearance of many leaves!

Conifers are trees that bear cones. Most conifers produce male and female cones.

Sometimes trees produce only male parts or only female parts, such as the yew.

The male cones are usually smaller than the female cones. The male cones contain pollen. They are usually found in clusters on the lower branches of the trees.

FASCINATING FACTS

Conifers produce an antifreeze like sap that allows it to carry nutrients even in freezing temperatures. This special "antifreeze" is what gives the conifer its strong piney smell.

FASCINATING FACTS

In damp weather, the cones of a fir tree close their scales tightly to protect the seeds. When the air dries, the scales pop open to release their seeds. Sometimes, this makes a loud, cracking noise.

The largest and most common order of gymnosperms is the conifer. Most conifers continue to make food all year long. They are called evergreens because they stay green all year.

Conifers include pines, redwoods, cedars, hemlocks, junipers, cypress, fir, and spruce.

sequoia   red cedar   redwood

Conifers are referred to as softwood trees because their wood is softer than the wood of oak, walnut, pecan, and maple trees.

The world's largest living thing is a sequoia tree in California named General Sherman. Its trunk is 80 feet around, the base is 36 feet across, and it is more than 272 feet high. It is thought to be about 3,500 years old. That means it started growing when the Egyptians were building their pyramids.

The leaves of conifers are usually needlelike or scalelike and vary in shape. Pines, cedars, and larches have needles. Yews, firs, and some redwoods have tough, flat, leathery leaves. Cypress trees have scales.

FF
FASCINATING FACTS

Cones of the sugar pine trees can grow to about 26 inches long.

FF
FASCINATING FACTS

The cones on a spruce tree point downward; the cones of a fir tree point upward.

The female cones contain seeds and are located higher up in the tree. In the spring, the female cones spread their scales to receive the pollen from the male cones.

Each female cone contains at least one ovule. The ovule contains the egg cell, which after fertilization, develops into a seed.

Wind blows pollen from the male cones to female cones. Seeds begin to develop. The female cone's scales close to protect the developing seeds. When the seeds ripen, the scales open up and the seeds fall out.

About 20% of the seeds dropped will germinate and only a small percentage of these seedlings will survive the first summer.

FF
FASCINATING FACTS

Coniferous trees supply about 75% of the world's timber and most of the world's paper.

FF
FASCINATING FACTS

Giant sequoias can get as big as 36 feet around – wide enough to drive a car through.

Throughout the year a conifer's old leaves drop off and are replaced by new ones. Since the leaves do not all fall at one time, most conifers are never leafless. Exceptions include the larch and bald cypress. These conifers shed their leaves in the fall.

Pine tree needles grow in bundles of two, three, or five.

Scotch pine needles are in bundles of twos. Monterey pine needles are in bundles of threes. Arolla pine needles are in bundles of fives.

Seeds are inside dry fruit like nuts and inside fleshy fruit like peaches. Plants produce seeds, and some of these seeds grow into plants that produce more seeds.

The seed coat protects a seed from animals, bacteria, extreme temperatures, and moisture loss.

A flower bed is full of plants called angiosperms.

Plants you see with flowers are angiosperms.

—cosmos
—petunias
—pansies

About 70% of the food we eat comes from seeds called grains. Rice, corn, rye, and wheat are seeds.

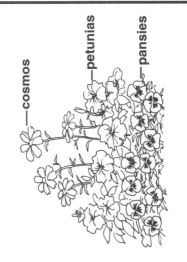

The seed of an angiosperm is enclosed in a case. Angio is a Greek word that means "covered" and sperm means "seed."

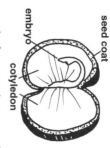

The seed of an angiosperm is enclosed in an ovary, which later becomes the fruit.

embryo · seed coat · cotyledon

WHO'S WHO

"Though I do not believe that a plant will spring up where no seed has been, I have great faith in a seed. Convince me that you have a seed there, and I am prepared to expect wonders." Henry David Thoreau

---

How do you think the seed coat protects the seed?

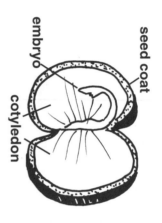

seed coat · embryo · cotyledon

The seed is made up of a seed coat, a tiny plant, and stored food.

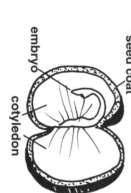

seed coat · embryo · cotyledon

The seed coat is called the testa. The tiny plant is called an embryo. The stored food is called the cotyledon.

Plants that produce flowers and seeds are called angiosperms.

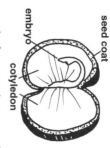

Angiosperms make up the largest class of plants with more than 240,000 species.

Angiosperms have

flowers,

leaves,

stems,

and

root systems.

Some seeds are eaten as snacks such as sunflower seeds and pumpkin seeds. Some seeds are used as spices, such as cumin and anise seeds.

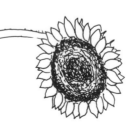

The stored food, or cotyledon, nourishes the tiny plant, or embryo.

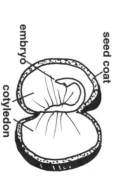

seed coat · embryo · cotyledon

The cotyledon nourishes the seed leaves until the true leaves appear. The true leaves then produce their own food through photosynthesis.

3. Seeds enclosed in fruit may be eaten by animals and passed on as waste.
4. A coconut is an example of a seed (within a fruit) scattered by water.
5. Some seeds that are scattered by exploding seed pods are impatiens and wisteria.

5

Seeds are scattered to allow a better chance for survival. They will have more room to grow and more nutrients available if they are not competing with the parent plant.

7

Some seeds may germinate immediately after they absorb water. Other seeds may remain "asleep" or unchanged until the conditions around them are suitable for germination.

This 'sleep time' is called dormancy. A seed may remain dormant for many years.

12   Lots of Science Library Book #13

When a seed begins to grow, the process is called germination. For a seed to germinate, it must have water, air, and suitable temperature. Many seeds require a period of cold weather before they can germinate.

Some seeds require a chemical change to occur, therefore depend on being passed through an animal's digestive system. Other seeds require extreme heat and rely on forest fires. Have you ever seen a controlled burning by foresters?

10   Lots of Science Library Book #13

Angiosperms begin life in seeds.

1

Lots of Science Library Book #13

Seeds fall from the parent plant and are scattered by wind, animals, and water. Seeds are moved in five main ways.

3

---

16   Lots of Science Library Book #13

FASCINATING FACTS

The seed of an Oriental lotus germinated after 3,000 years of dormancy.

14   Lots of Science Library Book #13

FASCINATING FACTS

Squirrels dig holes to store seeds and acorns. The ones the squirrels don't dig up to eat often germinate in the spring.

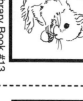

FASCINATING FACTS

The wild cucumber produces prickly fruit that grows bigger and juicier as it matures. The fruit holding the seeds swells and explodes, shooting the seeds 20 feet away from the parent plant at 60 MPH.

FASCINATING FACTS

An oak tree may produce 50,000 acorns, but only a few become trees. Explain why this happens.

---

1. Seeds scattered by wind are light and feathery such as dandelion seeds or streamlined like maple seeds.
2. Seeds scattered by animals are sticky or barbed such as grass seeds.

When an angiosperm is mature, or fully grown, it produces seeds.

FASCINATING FACTS

When birds eat elderberries, the seeds are passed through their droppings. This moves the seeds and helps them germinate.

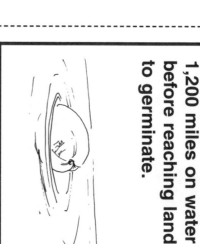

FASCINATING FACTS

A coconut may travel 1,200 miles on water before reaching land to germinate.

It is a monocot.
Mono is a Greek prefix
meaning "one."

cotyledon

Most of the seed plants
you are familiar with are
dicot plants. There are
many more dicot plants
than monocot plants in
the world.

WHO'S· WHO

The Greek
philosopher,
Theophrastus,
(around 300 B.C.) is
credited with being
the first to recognize
that there were two
types of seeds,
monocot and dicot.

Much of our food
source comes from
monocot seeds such
as rice, rye, wheat,
and corn.

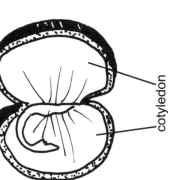

Let's review the parts of
a seed.

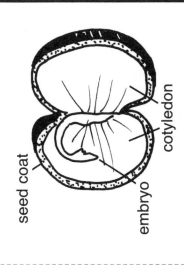

seed coat

embryo

cotyledon

Seeds with two
cotyledons are called
dicots. Di is a Greek
prefix meaning "two."

cotyledon

FASCINATING FACTS

Wheat germ comes
from the wheat seed
and is highly
nutritious.

Some monocots are lilies, tulips, trillium, palms, and grasses.

Lots of Science Library Book #14

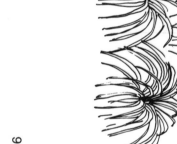

How are the lima bean and corn seed alike? How do they differ?

Lots of Science Library Book #14

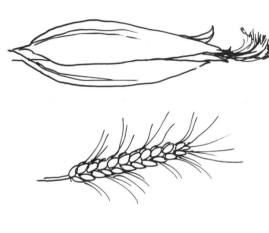

Here is a picture of the inside of a lima bean. How many cotyledons does it have?

seed coat

embryo

cotyledon

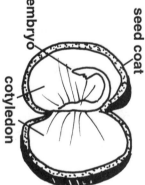

Lots of Science Library Book #14

There is a second type of seed with only one cotyledon. Look at the corn seed.

embryo

cotyledon

seed coat

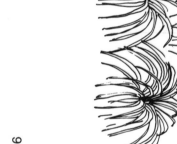

The embryos of the monocot and dicot receive nourishment from the cotyledons.

FASCINATING FACTS

The largest seed (fruit) comes from the coco de mer palm in the Seychelles Islands. The seed may grow as big as a beach ball and weigh more than 50 pounds.

There are two different types of root systems. One type of root has many roots that develop from the stem. This is called a fibrous root system. Monocot plants usually have fibrous root systems.

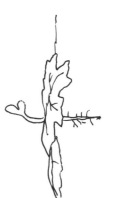

FASCINATING FACTS

The wild potato has a taproot that can weigh up to 30 pounds. It was an important food source for the early inhabitants of North America.

The root cap covers the tip of the root to protect it.

The root cap is slimy so it is easier to push through the soil. The cap is made up of dead cells.

12   Lots of Science Library Book #15

Roots always grow towards the center of Earth.

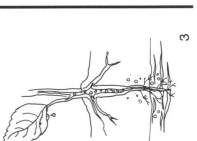

This is called geotropism and is the plant's natural response to gravity.

10   Lots of Science Library Book #15

What prevents the trees and plants from blowing away on a windy day?

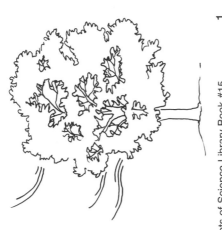

1

Lots of Science Library Book #15

Besides anchoring the plant, roots absorb water and minerals and store food.

3

FASCINATING FACTS

The branches of a Banyan tree grow down to the ground and roots begin new trunks.

16   Lots of Science Library Book #15

Roots form a passageway for water and minerals to be carried to the stem and leaves. Food made in leaves travels down the stem to the roots and other plant parts.

The passageway that carries water and minerals up from the roots to the rest of the plant is called the xylem. The passageway that carries the food down from the leaves to the roots is called the phloem.

14   Lots of Science Library Book #15

The second kind of root system grows a main root vertically down from the stem. More roots branch outwards and downwards. This is called a taproot system. Dicot plants usually have taproot systems.

main root

Carrots, beets, and radishes are examples of taproots. A taproot has a root cap and root hairs.

The epidermis is the outer covering that protects the inner tissues of the root.

FASCINATING FACTS

Most trees cannot grow where the ground is constantly in water because their roots need air. Plants like mangroves and bald cypress have "breathing roots" which are exposed to the air so they can get the oxygen they need to survive. Breathing roots are called pneumatophores.

The root emerging from the embryo is called the radicle.

As germination begins, the first thing to emerge from the seed is a root.

Roots anchor plants to the ground.

FASCINATING FACTS

The first plant to flower in space was grown aboard the Soviet Union's Salyut-7 space station in 1982. Since there is no gravity in space, how do you think the roots looked?

Tiny root hairs help the plant take in more water and minerals.

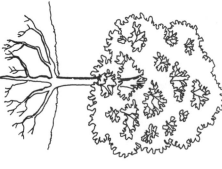

Stems also store food.

5

Dicot and monocot plant stems do the same things, but differ in a few of their characteristics.

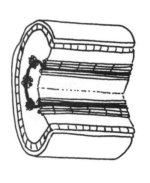

When you cut a flower to put it in a vase, you usually cut it with a long stem attached. The flower needs the stem to continue receiving water.

1

Stems can be hard and rigid. Hard stemmed plants include maple trees and rose bushes.

Hard stemmed plants are called woody plants.

16   Lots of Science Library Book #16

Stems are covered with a thick outer layer to prevent water from evaporating and to protect the plant from disease.

The protective layer is called the epidermis.

7

Both of these passageways make up the vascular bundles.

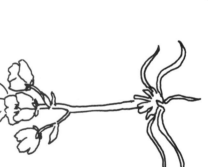

The stem of a plant supports the leaves and the flowers.

3

The stems of monocot plants contain a limited number of vascular bundles. They are arranged in random order. After monocot plants reach maturity, the stem does not thicken.

14   Lots of Science Library Book #16

These passageways going up are called xylem.

Passageways inside the stem carry water and minerals from the roots to the other plant parts.

FF

FASCINATING FACTS

Potatoes, sweet potatoes, and yams are not roots, but tubers, the fleshy part of an underground stem.

These passageways going down are called phloem.

Other types of passageways carry food made in leaves down to the roots.

FF

FASCINATING FACTS

When a maple tree is tapped for syrup, the sap is boiled to remove the water so that only the sugary phloem sap remains.

The sap is made up of the fluids of the xylem and phloem.

Cactus stems are thick and fleshy and are used to store water.

The stem carries water and food throughout the plant.

Dicot stems contain a large number of vascular bundles. The vascular bundles are arranged in a ring inside the stem. As a dicot plant grows, its stem grows thicker too, throughout its life.

Stems usually grow above the ground towards the light.

The stem's positive response to the light is called phototropism.
See *Lots of Science Library Book #4.*

Some stems are soft and greenish in color. Examples of soft stemmed plants are dandelions, tomatoes, zinnias, sunflowers, peas, and grass.

Soft stemmed plants are called herbaceous plants.

The leaf blade is the main, flat part of the leaf.

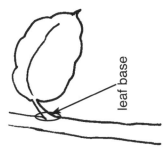

blade (lamina)

5

The leaf base attaches the leaf stalk to the stem.

leaf base

7

Food and water are carried throughout the leaves by veins. The veins also strengthen the leaf's structure.

The xylem carries water and minerals up to the leaves. The phloem carries the food made in the leaves to the other parts of the plant.

12   Lots of Science Library Book #17

Inside the leaf, the plant produces its own food and exchanges gases.

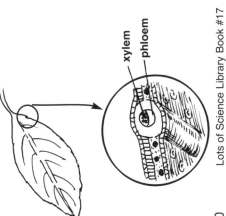

xylem
phloem

10   Lots of Science Library Book #17

Do you remember the important functions of leaves?

Leaves capture sunlight and make their own food through photosynthesis.

1   Lots of Science Library Book #17

The leaf has three parts: leaf blade, leaf stalk, and leaf base.

The leaf blade is called the lamina. The leaf stalk is called the petiole.

3

FASCINATING FACTS

The largest leaves are from the raffia palms of Africa and Madagascar. They can grow to be more than 65 feet long!

16   Lots of Science Library Book #17

Veins of monocot and dicot plants do the same job but are different. The veins of a monocot plant, like a tulip, run parallel, or side-by-side.

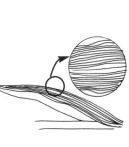

14   Lots of Science Library Book #17

The leaf stalk is the part that connects the leaf blade to the stem. The leaf stalk can bend easily so that leaves are not broken off by the wind.

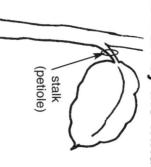

stalk
(petiole)

Lots of Science Library Book #17

The leaf is covered with a thin outer layer to protect it from:
1. disease
2. bacteria
3. water loss

cuticle
epidermis

The protective covering is called the epidermis. The epidermis is covered with a thin layer called the cuticle. The cuticle gives the leaves their waxy appearance.

Lots of Science Library Book #17

FASCINATING FACTS

The leaf of the royal waterlily in Brazil can grow to eight feet in diameter and hold up to fifty pounds.

FASCINATING FACTS

The spines of a cactus are actually leaves. The tendrils of vines are leaves that act like stems.

FASCINATING FACTS

Man-eating plants! There is no such thing. There are plants that capture insects with their leaves and digest them, such as the pitcher plant, sundew, and Venus flytrap.

Lots of Science Library Book #17

FASCINATING FACTS

Some leaves are poisonous. Caterpillars of the monarch butterfly eat poisonous milkweed leaves. After eating these leaves, the butterflies become poisonous, too. This protects them from being eaten.

The veins of a dicot leaf, like a hibiscus, are netlike.

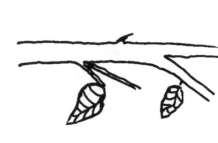

FASCINATING FACTS

Banana trees are not really trees at all. Their stalks are made of tightly coiled leaves. They can grow to about 30 feet tall.

A flower is made up of four rings of leaves.

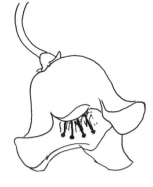

The rings of leaves are called whorls. Flowers with all four whorls are called complete flowers. Flowers lacking any one of the whorls are called incomplete flowers.

The second ring is made up of petals. These are usually the showy, colorful part of the flower. The petals attract insects and birds. They protect the male and female parts of the flower.

←petal

The flower head is surrounded by many small flowers called ray florets.

Although a thistle consists of only disk florets it is considered a composite flower.

Look at the sunflower. How many flowers do you see? The flower head is made up of a cluster of many small flowers.

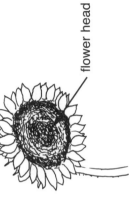

flower head

These small flowers are called disk florets.

When you think of a flower, what comes to your mind?

FASCINATING FACTS

The Puya raimondii grows in the Andes. It does not flower until the plant is about 150 years old. After it flowers, the plant dies.

FASCINATING FACTS

Not all flowers smell good. The rafflesia of Southeast Asia smells like rotting meat and is pollinated by flies. The rafflesia produces the world's largest flower. Most of the plant grows underground, but its giant red flower grows above the ground. Rafflesia's smelly red flower grows to about 4 feet across and can weigh up to 25 pounds.

Flowers of monocot plants and dicot plants do the same job but differ in the number of flower parts. Monocot plants have flower rings in multiples of three. Look at the lily. How many petals do you see?

The outer ring protects the bud and holds the flower in place.

The outer ring is called the sepal.

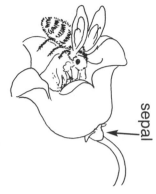

sepal

6

The many small flowers in the head are called disk florets.

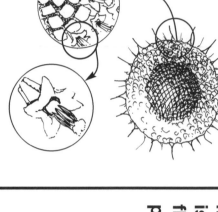

11

We enjoy looking at a flower and smelling its sweet smell. But the main purpose of a flower is reproduction. Without flowers, plants cannot produce seeds.

2

Dicot plants have flower rings in multiples of four or five. Look at the hibiscus. How many petals do you see?

15

8

The third ring is made up of male reproductive parts of the flower. A male reproductive part is called a stamen.

stamen

The fourth ring is made up of female reproductive parts of the flower. A female reproductive part is called a pistil.

pistil

9

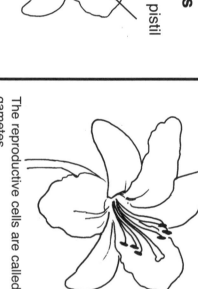

A flower contains the male and female reproductive parts.

The reproductive cells are called gametes.

4

Almost all dicot flowering plants are composite flowers, such as coneflower, daisy, dandelion, marigold, chrysanthemum, and zinnia.

Composite flowers make up one of the largest groups of flowering plants.

13

A stamen is the male reproductive part of the plant.

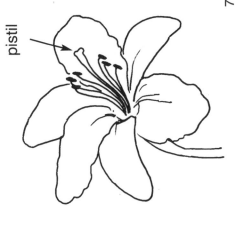

stamen

5

A pistil is the female reproductive part of the plant.

pistil

7

Pollen is passed from flower to flower by insects, birds, bats, and wind.

12   Lots of Science Library Book #19

As the animal visits another flower of the same kind, this pollen falls onto the sticky tip of the pistil.

10   Lots of Science Library Book #19

Do you remember the main purpose of the flower? Flowers are the reproductive parts of plants. They contain the male and female reproductive parts.

1   Lots of Science Library Book #19

The nectar, colorful petals, and sweet smell of the flower attract bees, butterflies, hummingbirds, and other animals.

3

FASCINATING FACTS

The black seeds in a watermelon are the pollinated seeds; the white seeds have not been pollinated. The more black seeds a watermelon contains, the tastier it is.

16   Lots of Science Library Book #19

When the seeds mature, they are scattered by wind, animals, water, or explosion.

14   Lots of Science Library Book #19

The stamen produces pollen, which contains male reproductive cells.

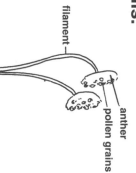

filament — anther — pollen grains

A stamen consists of a long slender filament supporting an anther at its end. The anther produces the pollen grains.

6

The pistil holds the immature seeds.

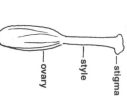

stigma — style — ovary

At the base of the pistil is the ovary, which contains the ovules, or immature seeds. This is connected to a long, slender tube called the style. At the tip of the style is the stigma, which is sticky and traps the pollen. Some flowers have one pistil, others have two or more.

8

The pollen absorbs moisture and grows a tube down to the hollow base of the pistil where seeds are produced. This entire process is called pollination.

The seeds are produced in the ovary.

11

When an animal enters a flower to drink its nectar, pollen from the flower's stamen collects on the animal's body.

9

Inside flowers is a sugar rich food called nectar. Nectar is the sweet liquid that bees collect to make honey.

2

FASCINATING FACTS

Some flowers have markings on the petals to help guide the insect to the center of the flower. Some of the markings can be seen only in ultraviolet light, so they are invisible to humans but clearly visible to bees.

4

FASCINATING FACTS

Plants that depend on the wind for pollen dispersal do not need a sweet smell or colorful petals.

15

FASCINATING FACTS

The best pollinators are honeybees because they produce colonies that need to provide food for 15 to 20 thousand bees a day.

13

As an animal moves from flower to flower, the pollen on its body falls on the sticky top of the flower female part of the flower (pistil).

Inside the hollow base are immature seeds, or eggs.

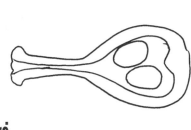

The immature seeds are called ovules.

Fruit trees such as apples, pears, peaches, plums, cherries, and apricots shed their leaves in the fall.

The purpose of the flower is to produce seeds. The purpose of the fruit is to scatter the seeds.

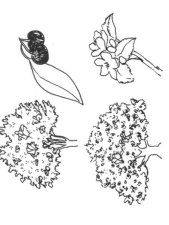

Let's review the part the flower plays in reproduction. The flower of a plant attracts bees and other animals.

FASCINATING FACTS

Navel oranges contain no seeds. Every navel orange you eat came from a single tree. In the early 1800's, an orange tree in Brazil produced an unusual orange without seeds. Buds from this tree were grafted onto another tree, buds from that new tree were grafted to another tree, and so on.

The pollen absorbs moisture and grows a tube down to the hollow base of the female part of the flower.

6

The hollow base of the pistil is called the ovary.

Other seeds are encased in a tasty flesh. This fleshy case is fruit you are most familiar with, such as apples, peaches, and plums.

11

Do you remember how seeds are scattered?

Seeds are scattered by wind, water, explosion, and animals. See *Lots of Science Library Book #13.*

2

Vegetables that produce seeds are really fruits. Common "fruit vegetables" are pumpkins, cucumbers, tomatoes, squash, okra, and eggplant.

15

After pollination, the ovary becomes fruit. Fruit develops around the seed.

Nuts are actually hard-shelled fruits. Their seeds are covered with a dry, tough fiber.

9

8

The fruit is the fertilized ovules and the ovary wall.

True nuts are hazelnuts, chestnuts, and filberts.

As an animal collects the nectar, pollen collects on its body. The animal visits other flowers of the same kind to collect more nectar.

4

Fruit trees such as grapefruit, lemons, oranges, and other citrus fruits stay green all year long.

13

Deciduous trees are also called hardwoods because their wood is hard. Do you remember which trees are softwoods? Conifers, which are gymnosperms, are softwoods.

5

Deciduous trees are sometimes called broad leaved trees because most of them have broad, flat leaves, unlike the needles of the conifers.

7

Deciduous trees with nonshowy flowers are wind pollinated.

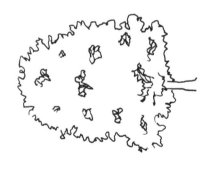

12

Deciduous trees produce flowers, although some of them are not colorful and showy.

10

Trees have woody stems and upper branches.

1

Angiosperm trees are also called deciduous trees.

3

FASCINATING FACTS
A tree's environment affects the shape of the tree. In windy places, trees may grow lopsided. Trees growing closely among other trees grow mostly upward to reach the sunlight. Trees with plenty of room to grow have a wide spreading crown.

16

FASCINATING FACTS
A golden delicious apple tree was sold in 1959 for $51,000, making it the most expensive fruit tree sold in the world.

14

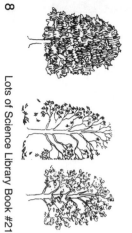

FF
FASCINATING FACTS

Forest fires can be helpful to trees. Fire burns ground clutter, giving seedlings a better chance for survival. Fire also helps open cones and seeds, thus aiding in germination.

Deciduous forests are found in various parts of the world, but they are most common in warm climates. To survive the cold seasons, they shed their leaves and 'rest' until spring.

Some varieties of oak are evergreen.

Examples of deciduous trees include maple, birch, oak, beech, sycamore, aspen, and dogwood.

Deciduous trees with colorful, scented, showy flowers are pollinated by insects and other animals.

FF
FASCINATING FACTS

The bamboo is not a tree; it is a type of grass.

The main stem of a tree is called the trunk. The upper part of the tree is called the crown.

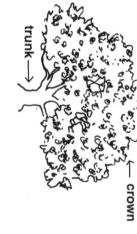

trunk →
← crown

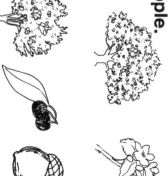

Tree flowers produce seeds. The seeds are often enclosed in a hard case like a nut or in a fleshy fruit like an apple.

Coniferous trees grow upward, while most deciduous trees grow by spreading their branches outward.

FF FASCINATING FACTS

The bark of seguoias can grow to be over two feet thick. The bark lacks resin which prevents it from burning as readily as other conifers.

FF FASCINATING FACTS

Cork comes from the bark of the cork oak tree. A layer of the outer bark is cut away every eight to ten years.

The tree's roots grow more outward than downward. They may spread as far outward as the tree's height.

FF FASCINATING FACTS

Trees can be hollow inside and yet remain healthy. A tree can live without its heartwood.

heartwood

A tree has a woody stem. The main stem is called the trunk. Look at the parts of the trunk of a tree.

bark

heartwood

cambrium

FF FASCINATING FACTS

Cinnamon is a spice that comes from the bark of the cinnamon tree. Strips of bark are cut from young saplings. They curl as they dry, forming the cinnamon stick!

Millions of root hairs grow from the root. They absorb water and minerals from the soil. Root hairs live only a short time, but are quickly replaced by new ones.

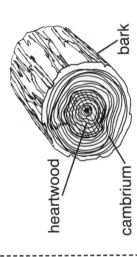

A close look at the bark of a tree reveals cracks. These cracks allow gas exchange between the air and the inside of the trunk.

These cracks are called fissures.

Under the bark is the cambium. This is the part that grows and thickens the tree.

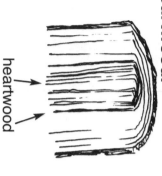

The cambium produces phloem cells on one side and xylem cells on the other. The cambium grows a layer of wood every year, producing the tree's rings.

A tree's trunk contains vascular tissues to transport water and nutrients from the root system to the other tree parts.
The trunk also moves food produced in its leaves to the root system.

Each year as new layers form, the older layers die. This results in a dead core that gives the tree strength. The layer of dead cells is called heartwood.

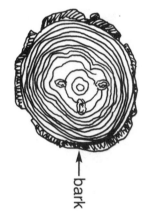

heartwood

The outer covering, or bark of a tree is made up of dead cells. These cells protect the tree from extreme temperatures, insects, and fungi.

bark

Bark is very thin on some trees and thick on others.
Birch trees have thin bark, about one inch thick.
Douglas fir trees have bark about one foot thick.

The young roots absorb the nutrients and the old roots anchor the tree.

Root systems of trees hold soil in place. This function helps the tree stay anchored and benefits nearby animals as well.

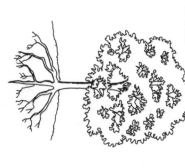

As summer ends and the weather begins to cool, tree growth slows down and trees form smaller thick-walled cells to preserve water.

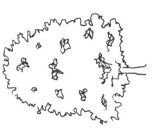

5

---

12   Lots of Science Library Book #23

In the fall, a tree's growth stops completely until next spring.

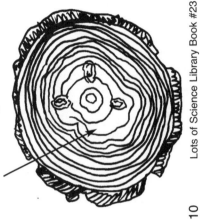

7

During a wet summer trees grow more, so there will be a wide ring.

wet summer rings

10   Lots of Science Library Book #23

Have you ever looked at the rings in a tree stump?

The rings form concentric circles.

Lots of Science Library Book #23   1

In spring trees grow quickly because the weather is warm and they receive plenty of rain.

The wood that forms during spring is made up of thin-walled cells.

3

---

16   Lots of Science Library Book #23

Tree rings reveal more than the age of a tree. They can tell scientists about rainfall, climate, and events such as fires, earthquakes, and pollution during the tree's lifetime.

Lots of Science Library Book #23   14

One pair of light and dark colored rings represents one year of tree growth.

These thick-walled cells form a dark colored wood ring in the trunk.

During a dry summer, trees do not grow very much, so the ring will be narrow.

The dark colored ring is called latewood.

latewood

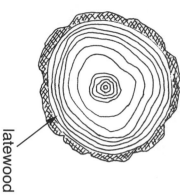

The newer rings grow outward and make the trunk thicker. Remember, a mature tree's trunk grows outward in diameter, not upward in height.

---

dry summer rings

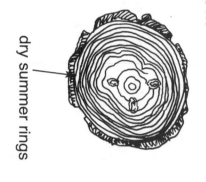

Look at the cross section of a tree trunk.

narrow rings

wide rings

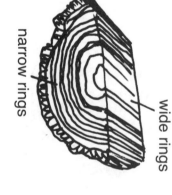

These thin-walled cells form a light colored wood ring.

The light colored ring is called earlywood.

earlywood

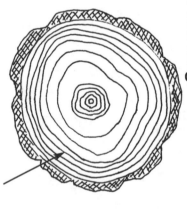

Weather is a major factor in determining the growth size of a tree's annual rings.

FASCINATING FACTS

You don't need to kill a tree to determine its age. Dendrochronologists use a tool called an increment borer to extract wood core samples the size of a drinking straw.

When the leaf blade is divided into separate parts, it is called a compound leaf.

Dicot plants have simple or compound leaves.

5

Just as we prepare for winter, trees do the same.

7

In the winter, when the ground is cold, trees cannot absorb much water.

12

FF

FASCINATING FACTS

There are hundreds of tiny holes on the underside of a leaf. This allows air to move in and water to move out of the leaf. These holes are called **stomata**.

10

Deciduous trees have different kinds of leaves.

1

16

When the leaf blade is one piece, it is called a simple leaf.

Most monocot plants have simple leaves.

3

Without water and light, the leaf cannot make food, so the tree's food factory shuts down. As the green chlorophyll breaks down other colors, such as yellow, red, and orange, show on the leaf.

14

In autumn, trees receive less sun and less rain. Their roots have a harder time getting the water they need.

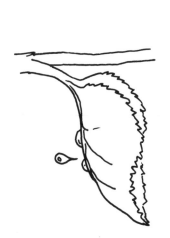

Lots of Science Library Book #24

Lots of Science Library Book #24

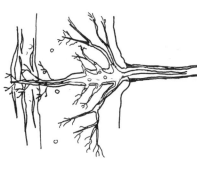

What do you do when the seasons change and warm summer breezes become cool autumn winds? You change into warmer clothes.

Do you remember that plants lose water through their leaves?

In spring and summer, roots usually find plenty of water to replenish the water lost through the leaves.

Look at these tree leaves. How are they similar? How do they differ?

Look at these leaves. How are they similar? How do they differ?

Lots of Science Library Book #24

Look at these tree leaves. How are they similar? How do they differ?

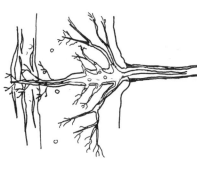

Lots of Science Library Book #24

Eventually, due to the lack of food and water, the tree's leaves fall to the ground.

To prevent the tree from losing all its water, the tree forms a layer of cork at the end of each leaf stalk, which stops the flow of water to the leaves.

Graphics Pages
and
Lab Log

Note: The owner of the book has permission to photocopy the *Lab Log* and *Graphics Pages* for his/her classroom use only.

Lab Log

Sunday	Monday	Tuesday	Wednesday	Thursday	Friday	Saturday

Investigative Loop ™

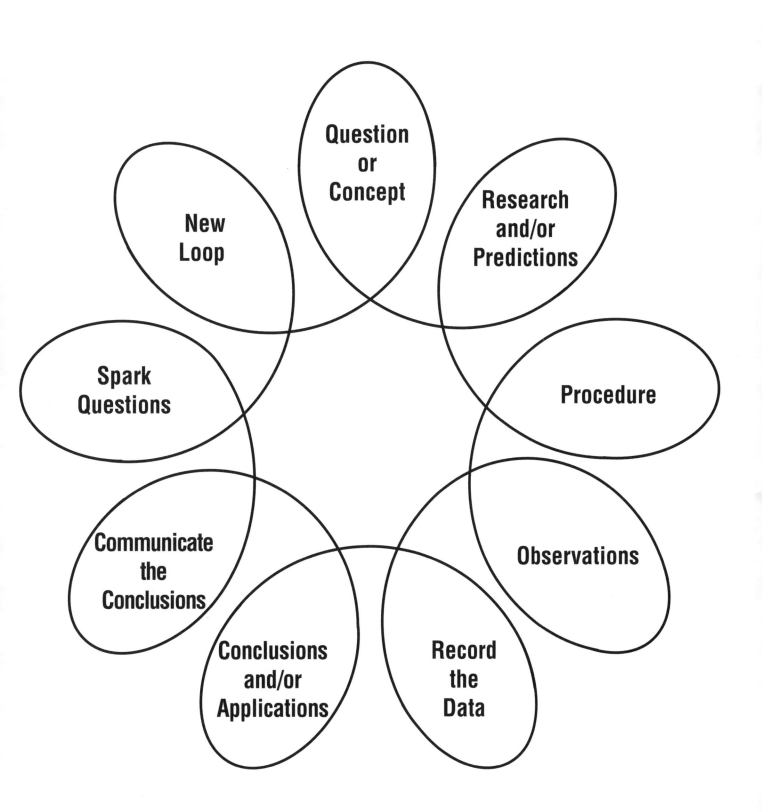

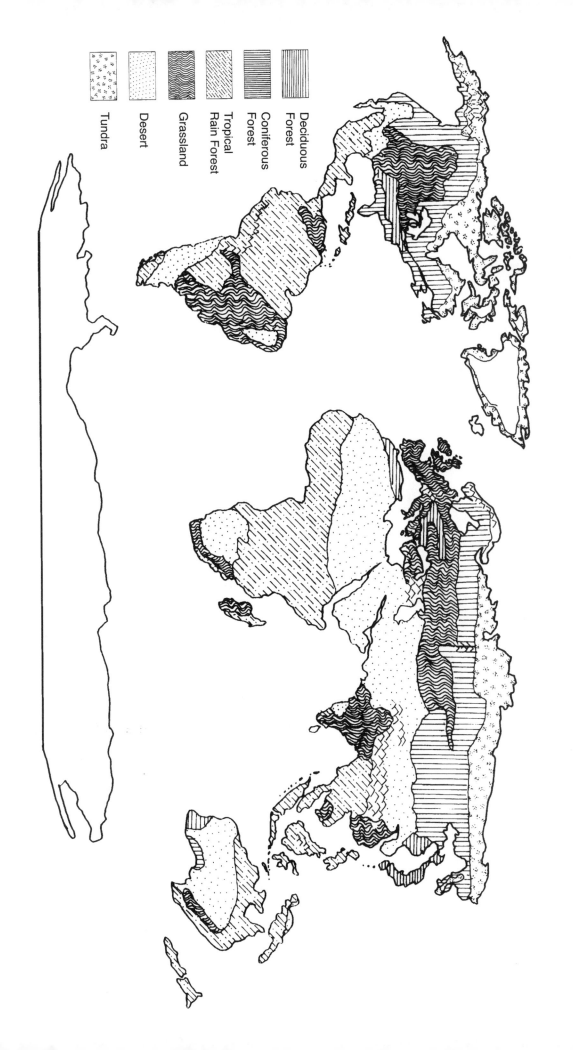

World Map of Forests

Tundra
Desert
Grassland
Tropical Rain Forest
Coniferous Forest
Deciduous Forest

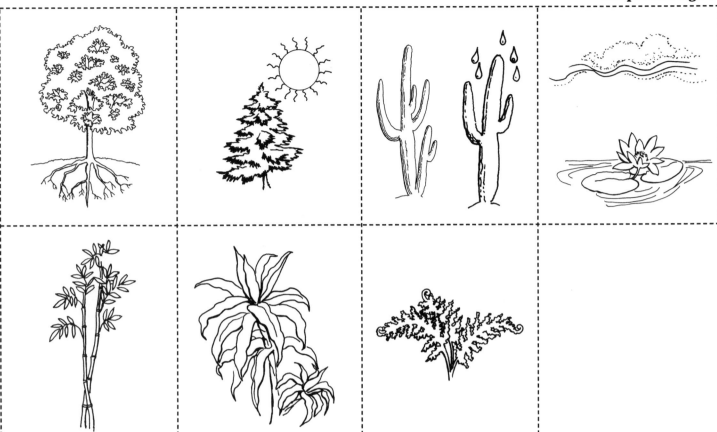

Photosynthesis   2-A   2-B   2-C   2-D

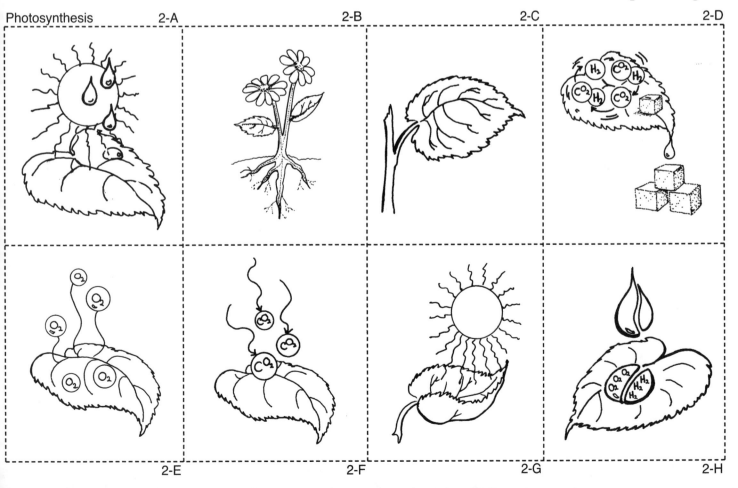

2-E   2-F   2-G   2-H

Chlorophyll in Plants Lab 2-1

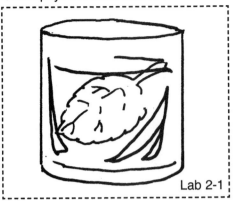

Lab 2-1

The Respiration Cycle 3-A

Leaves Give Off Oxygen Lab 3-1

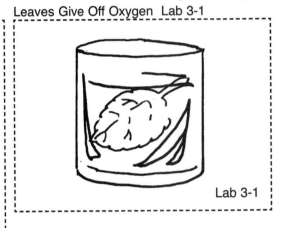

Lab 3-1

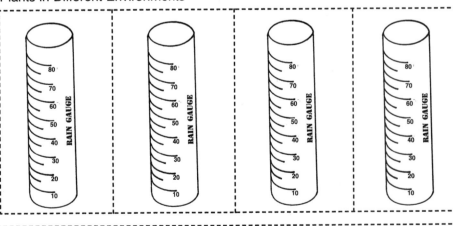

Leaves Transpire Lab 3-2

Lab 3-2

Plants in Different Environments

Plants in Different Environments (cont'd)

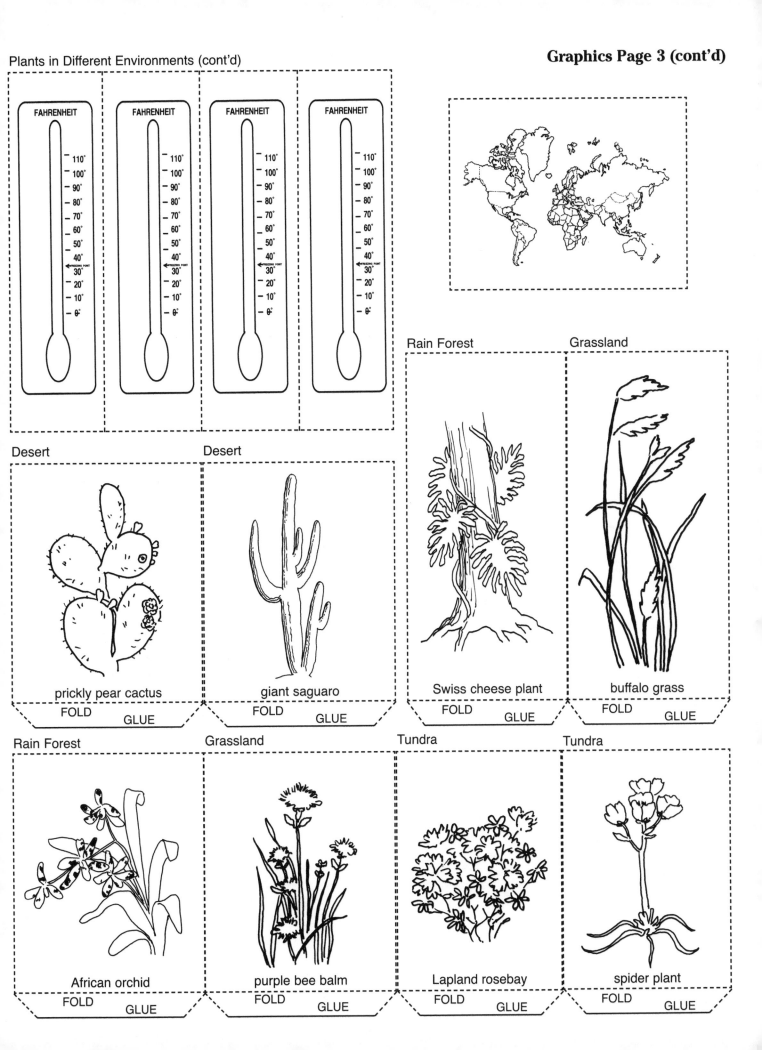

FAHRENHEIT
— 110°
— 100°
— 90°
— 80°
— 70°
— 60°
— 50°
— 40°
← FREEZING POINT
— 30°
— 20°
— 10°
— 0°

FAHRENHEIT
— 110°
— 100°
— 90°
— 80°
— 70°
— 60°
— 50°
— 40°
← FREEZING POINT
— 30°
— 20°
— 10°
— 0°

FAHRENHEIT
— 110°
— 100°
— 90°
— 80°
— 70°
— 60°
— 50°
— 40°
← FREEZING POINT
— 30°
— 20°
— 10°
— 0°

FAHRENHEIT
— 110°
— 100°
— 90°
— 80°
— 70°
— 60°
— 50°
— 40°
← FREEZING POINT
— 30°
— 20°
— 10°
— 0°

Desert

Desert

prickly pear cactus

FOLD   GLUE

giant saguaro

FOLD   GLUE

Rain Forest

Grassland

Swiss cheese plant

FOLD   GLUE

buffalo grass

FOLD   GLUE

Rain Forest

Grassland

Tundra

Tundra

African orchid

FOLD   GLUE

purple bee balm

FOLD   GLUE

Lapland rosebay

FOLD   GLUE

spider plant

FOLD   GLUE

Tropism and Sunshine Lab 4-1

Lab 4-1

Tropism and Water Lab 4-2

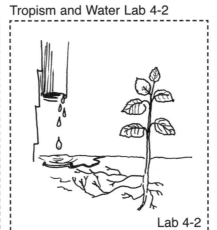

Lab 4-2

Asexual Reproduction Lab 5-1

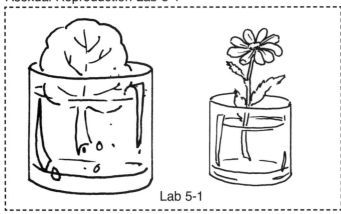

Lab 5-1

How Plants Reproduce 5-A

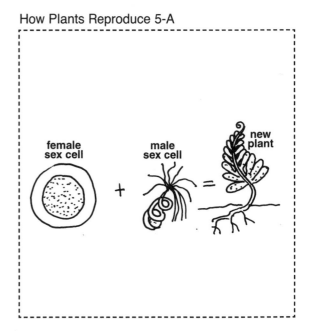

female sex cell + male sex cell = new plant

5-B

5-C

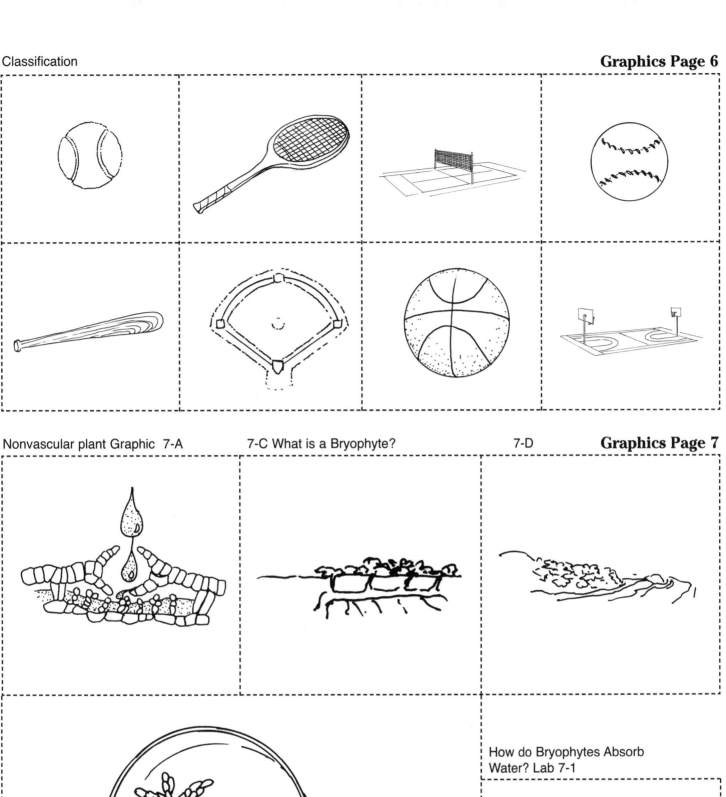

How do Bryophytes Absorb
Water? Lab 7-1

Lab 7-1

7-B

How do Bryophytes Reproduce?   8-A

8-B

8-C

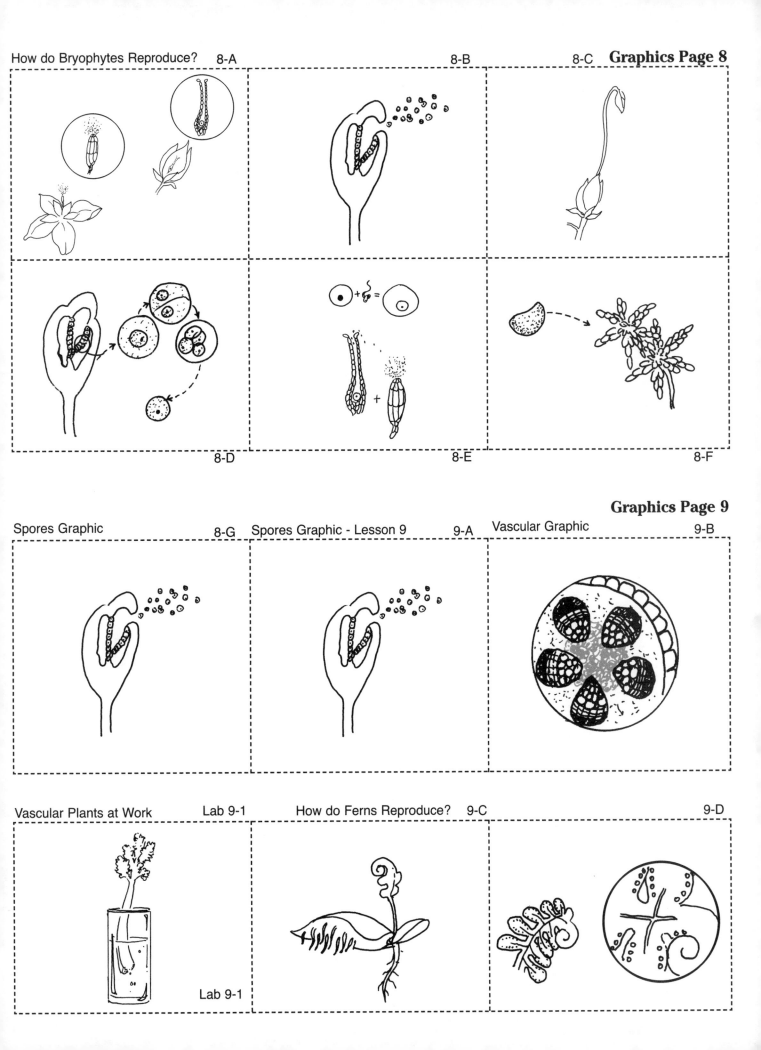

8-D

8-E

8-F

Spores Graphic   8-G

Spores Graphic - Lesson 9   9-A

Vascular Graphic   9-B

Vascular Plants at Work   Lab 9-1

Lab 9-1

How do Ferns Reproduce?   9-C

9-D

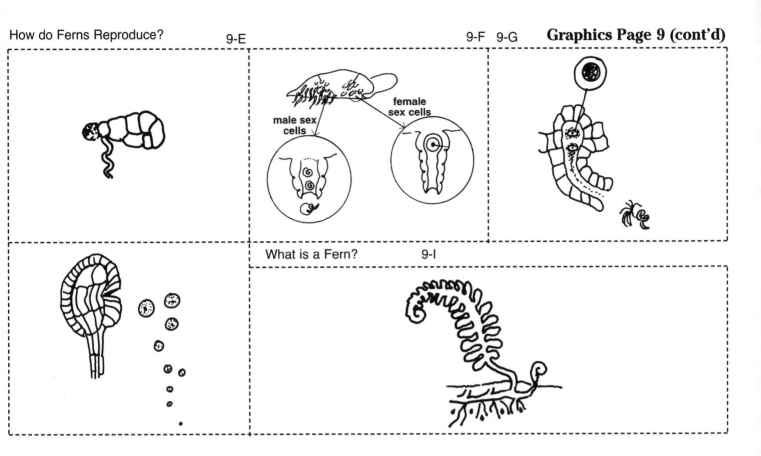

male sex cells

female sex cells

What is a Fern?   9-I

Vascular Graphic   10-A

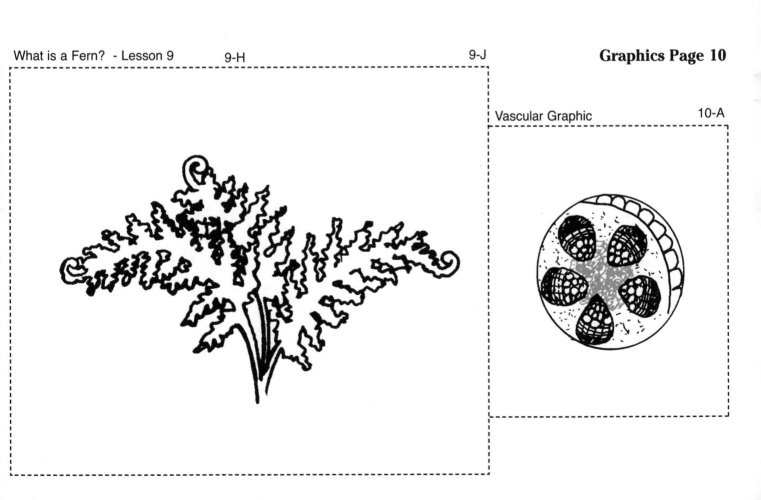

Conifers are Gymnosperms   11-A   11-B **Graphics Page 11**

Seed Graphic   11-C

Germination - Lesson 13   Lab 13-1

A Look Inside a Seed   Lab 12-1 Vascular Graphic   12-A

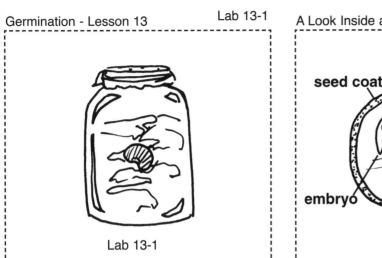

Lab 13-1

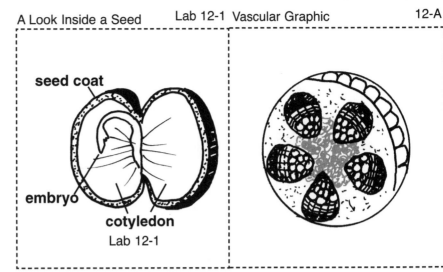

seed coat

embryo

cotyledon
Lab 12-1

How Seeds Travel, Dormancy, and Germination   13-A

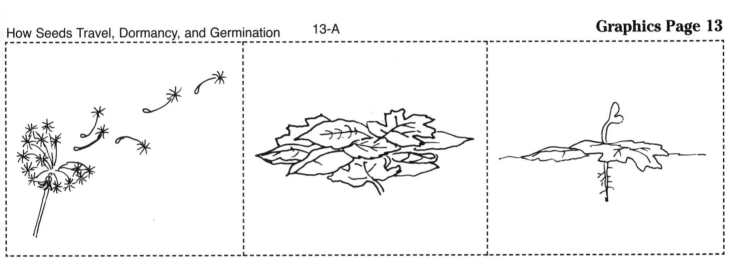

Seed Travelers   13-B

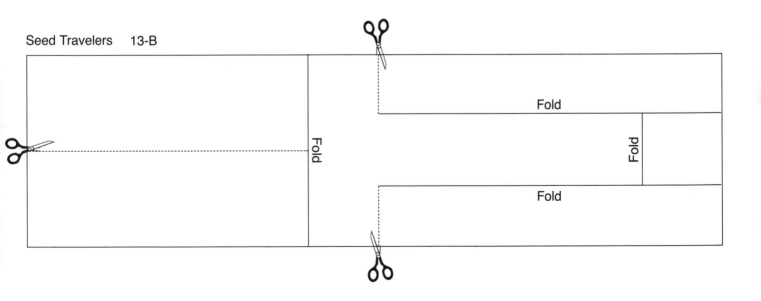

Fold

Fold

Fold

Fold

Monocot Seed and Dicot Seed

14-A

14-B

Monocot Seed and Dicot Seed   14-C

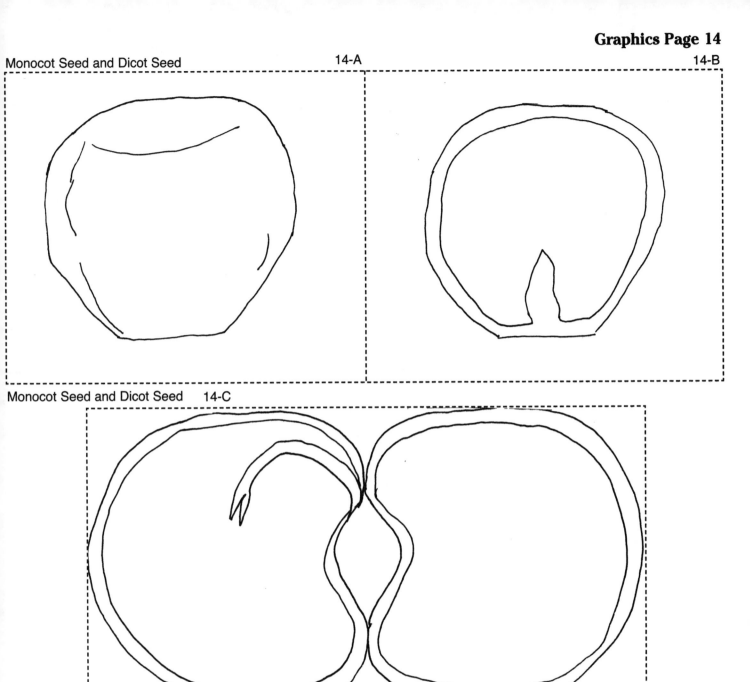

14-D

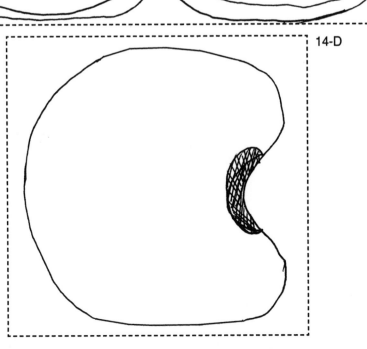

15-A Angiosperm Roots 15-B

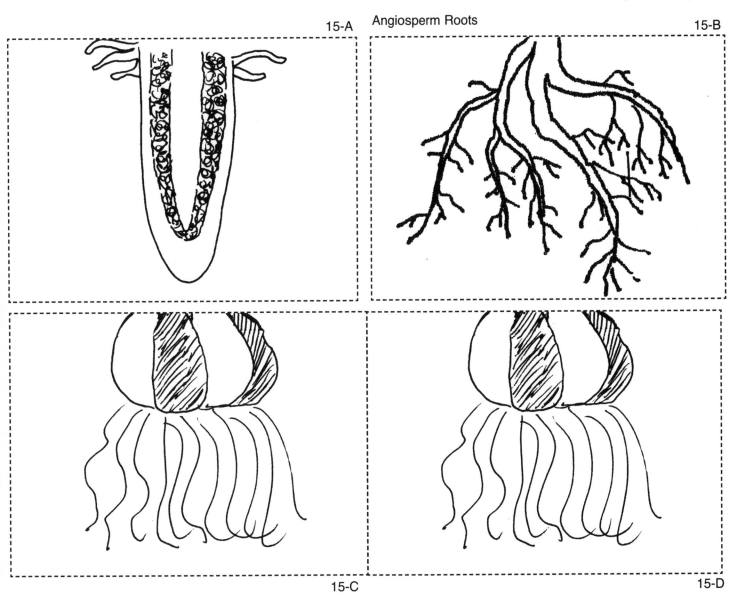

15-C 15-D

Functions of Roots 15-E

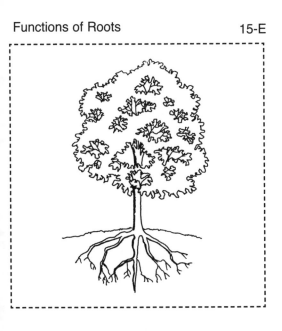

What Do Roots Do?   15-F

15-G

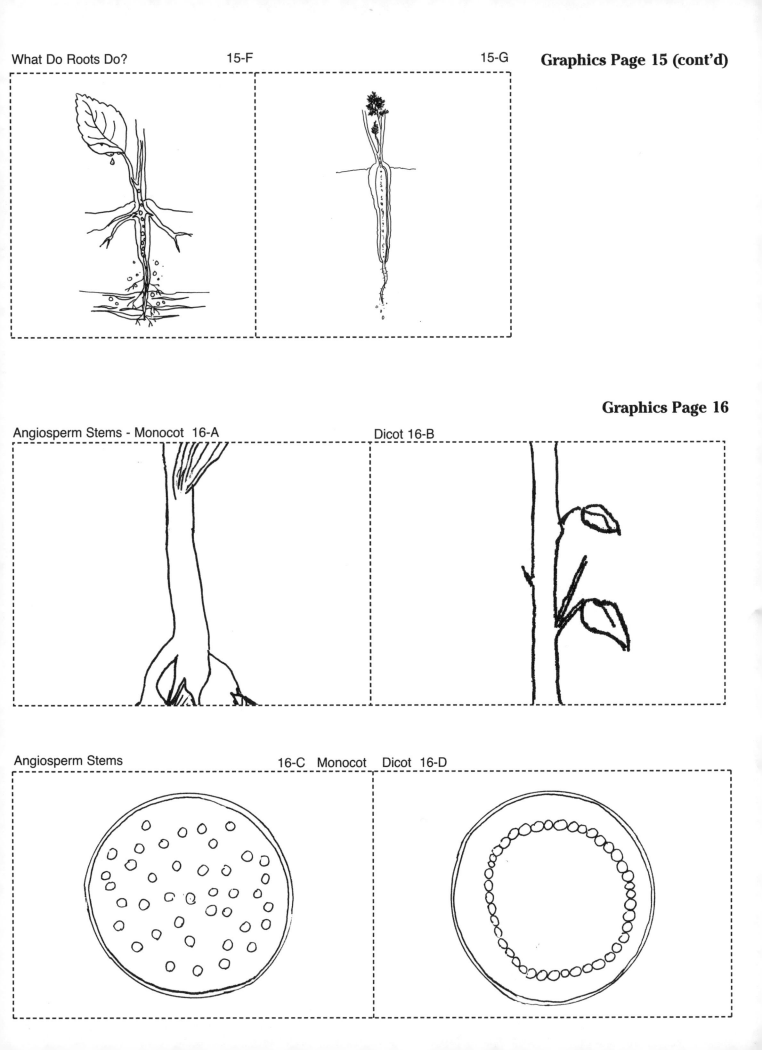

Angiosperm Stems - Monocot 16-A

Dicot 16-B

Angiosperm Stems   16-C Monocot   Dicot 16-D

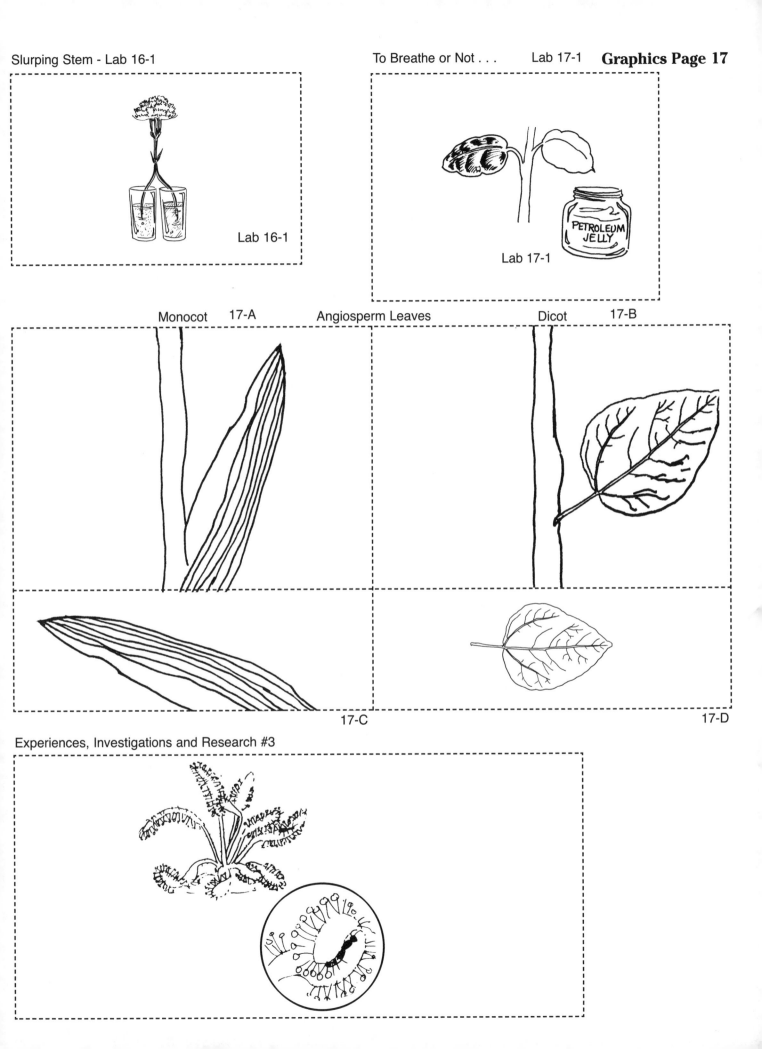

Slurping Stem - Lab 16-1

Lab 16-1

To Breathe or Not . . .   Lab 17-1   **Graphics Page 17**

PETROLEUM JELLY

Lab 17-1

Monocot   17-A   Angiosperm Leaves   Dicot   17-B

17-C

17-D

Experiences, Investigations and Research #3

Composite Flower   18-A

The Flower - Monocot   18-B   Dicot   18-C

The Flower - Inside Pictures   18-D   18-E

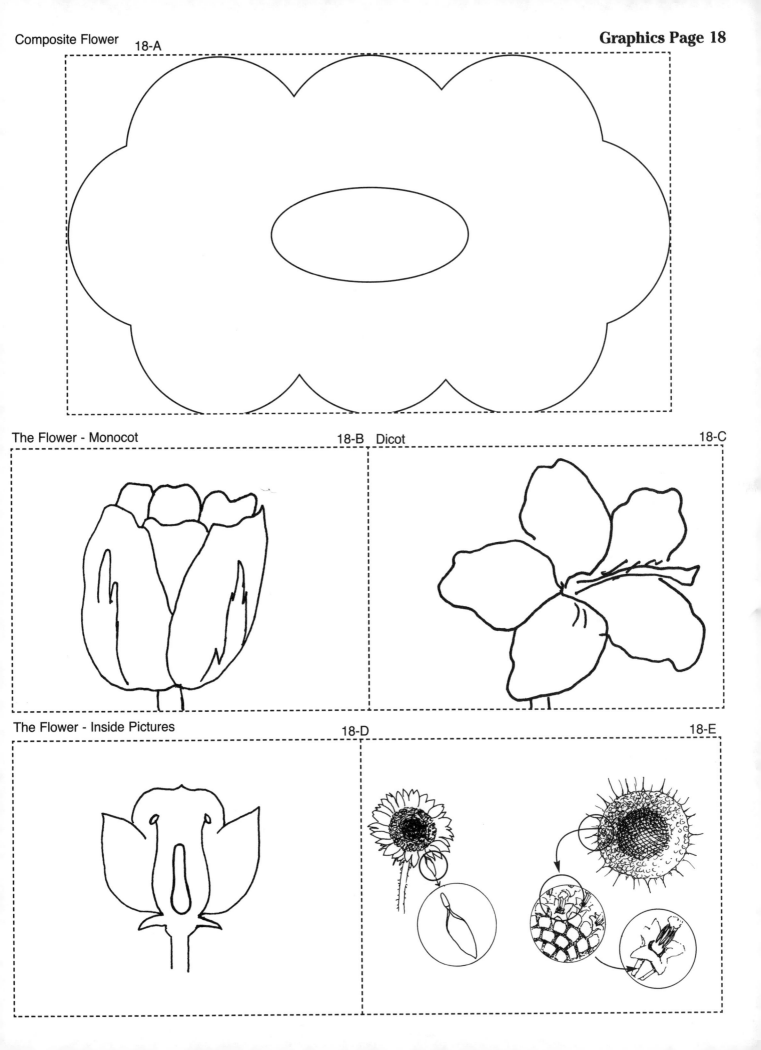

Seeds Graphic   19-E

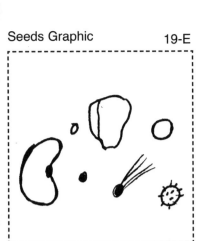

Germinating Pollen Grains Lab 19-1

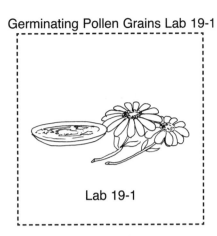

Lab 19-1

Pollination Cover

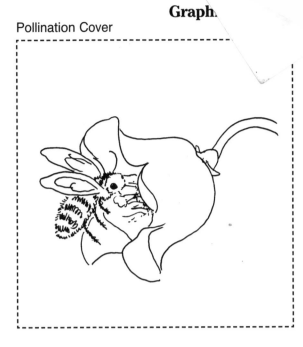

Flower Graphic   19-F

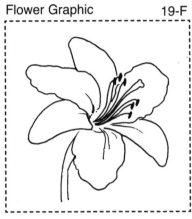

Pollination   19-A   19-B

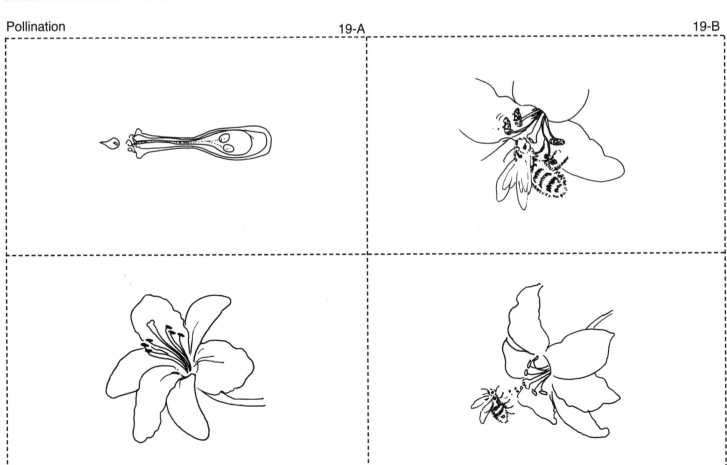

19-C   19-D

Coniferous and
Deciduous Trees

Cut #1

Cut #1

Coniferous and Deciduous Trees

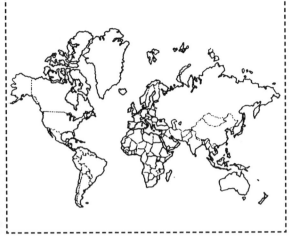

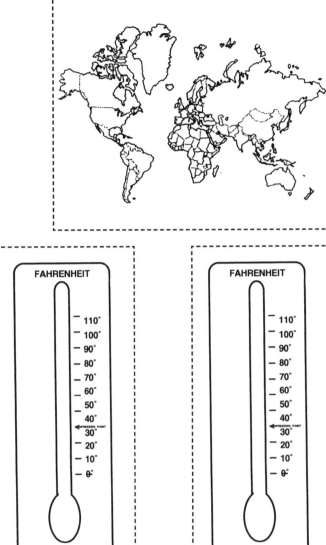

Coniferous and Deciduous Trees   22-A

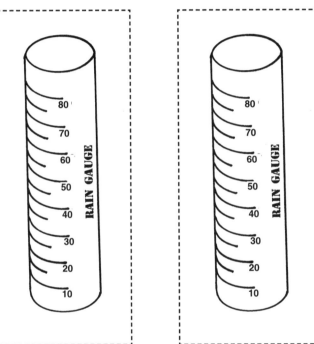

Tree Rings
23-A

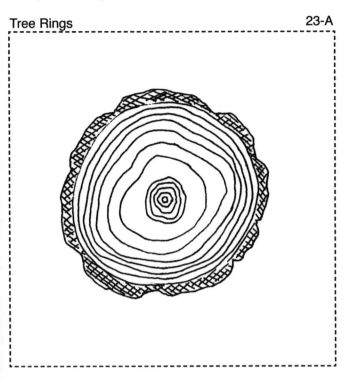

Trick a Leaf
Lab 24-1

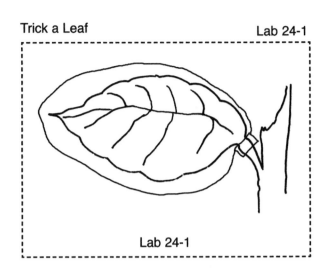

Lab 24-1

Leaves Change Colors
24-A

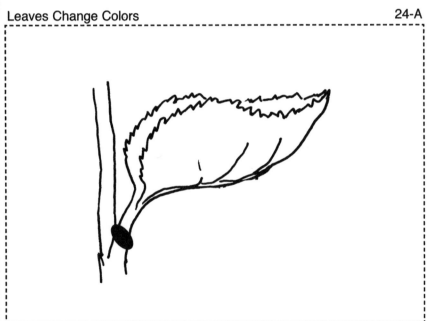

Sort Leaves
24-B

Sort Leaves
24-C

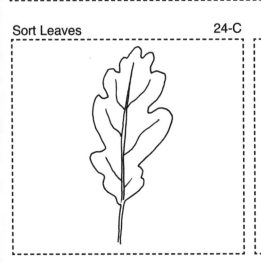

24-D

24-E

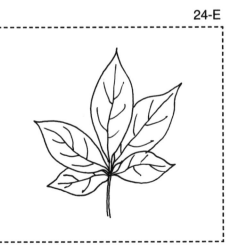

Seasons of a Deciduous Tree 24-F

24-G

24-H

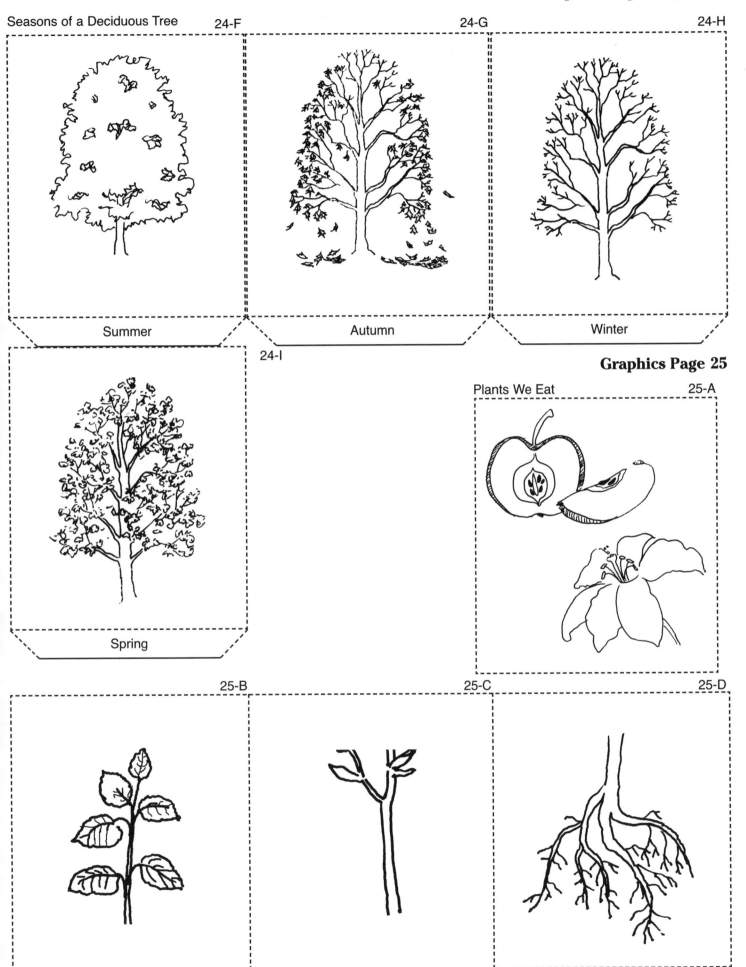

Summer

Autumn

Winter

24-I

Spring

Plants We Eat 25-A

25-B

25-C

25-D

Tons of Food   25-E

Tons of Food	Apples	Bananas	Carrots	Corn	Onions	Oranges	Peas	Potatoes	Rice	Strawberries	Tomatoes	Wheat
2 Million Tons												
1.5 Million Tons												
1 Million Tons												
500,000 Tons												
	148,000	159,000	43,000	1.8 Million	98,000	176,000	13,200	801,000	1.6 Million	6,600	234,000	1.5 Million